Pelican Books
Genetic Risk

Stephen Thomas, a freelance editor both of law books and of less specialised works, was born in 1951 and educated at Harrow, Oriel College, Oxford, where he read history, and the College of Law, London.

As a one-time consumer of genetic counselling, he has a keen interest in the evolution of medical genetics. His papers on the subject have been read to the British Psychological Society and published in the *British Medical Journal*.

Stephen Thomas

GENETIC RISK
A Book for Parents and Potential Parents

PENGUIN BOOKS

Penguin Books Ltd, Harmondsworth, Middlesex, England
Viking Penguin Inc., 40 West 23rd Street, New York, New York 10010, U.S.A.
Penguin Books Australia Ltd, Ringwood, Victoria, Australia
Penguin Books Canada Limited, 2801 John Street, Markham, Ontario, Canada L3R 1B4
Penguin Books (N.Z.) Ltd, 182–190 Wairau Road, Auckland 10, New Zealand

First published 1986

The illustration on page 60 appears in C. R. Austin and R. .V. Short (eds.),
Reproduction in Mammals, Book 2: *Embryonic and Fetal Development* (Cambridge, 1982)
and is reproduced by permission of Masson S.A., Paris, and Cambridge University Press.

Made and printed in Great Britain by
Richard Clay (The Chaucer Press) Ltd, Bungay, Suffolk
Typeset in Monophoto Bembo

To

Jane

Contents

Preface

A few years ago I visited Britain's oldest Medical Genetics Clinic in order to receive counselling about a mutant gene. The clinic wasn't exactly ancient. Nor was I. Like most people outside the medical profession, I had never really thought about the ways in which medical genetics is setting us apart from our ancestors. I knew very little about genes, chromosomes and reproductive risk. My ignorance on that wretched occasion made the counsellor's task difficult. And it dawned on me, later, that hospital clinics are not always the best places to *start* learning about genetic risk.

Not wanting to write about just one of the thousands of genetic disorders man is subject to, I have chosen to write about genetic risk in general. My book is not narrowly disease-specific.

Genetic Risk is intended primarily for parents and potential parents. But it draws attention to problems that affect us all. The simple illustrated text and glossary provide stepping-stones, designed to enable even unscientifically minded readers to cross part of the chasm that separates the intellectual territory of the expert medical geneticist from our own. The book can't turn anybody into a scientific expert but does give people a chance to get to know some of the ideas and special terms geneticists use. Should anyone who has read the book be referred for genetic counselling (or actively seek counselling for themselves), the counsellor won't have to start from scratch. There will be an area of common ground.

Coping with genetic risk is not solely a matter of the intellect, of course. Having understood the words of a genetic counsellor, we must respond. Many of our choices, many of our responsibilities, are new, as the historical bits of the book demonstrate.

I say a little about coming to terms with past genetic misfortune. But most of the book looks to the future, drawing attention to numerical risk estimates, carrier detection tests, prenatal diagnostic tests, selective abortion, contraception, sterilisation, donated genes and adoption.

It will be obvious that a short introduction to a vast subject cannot be authoritative. The labour of producing this book will have been rewarded if the reader turns to specialist works and takes individually tailored professional advice on the many matters I've glanced at.

I am indebted to the following people for information (the names of those who very kindly read parts of early drafts of this book are marked with an asterisk; I am very grateful to them and to my family for their comments and help): Rachael Astor (Department of Oriental Antiquities, British Museum); Professor D. T. Baird (Centre for Reproductive Biology, University of Edinburgh); Toni Belfield (Medical Information Officer, Family Planning Information Service); Dr Caroline Berry (Consultant in Clinical Genetics, Guy's Hospital); Hon. Fiona Campbell; the late Professor C. O. Carter* (formerly Director of the Medical Research Council's Clinical Genetics Unit, Institute of Child Health, London); Members of the Cystic Fibrosis Research Trust; Dr Audrey Eccles; Mrs Rosemary French; Professor P. S. Harper* (Consultant in Medical Genetics, University of Wales College of Medicine); Dr John Havard (General Secretary, the British Medical Association); Mrs M. A. Herron; Miss Patricia Gallagher (Branch Development Officer, The Society for the Protection of Unborn Children); Hilary King (Press Officer, the British Medical Association); Mr John McCracken (Department of Health and Social Security); Professor Ivana Markova* (Department of Psychology, University of Stirling); Theresa-Mary Morton (Photographic Department, Royal Library, Windsor); Mr Oliver Parker* (English Law Commission); Dr M. A. Patton* (Hon. Senior Registrar in Clinical Genetics, Great Ormond Street Children's Hospital); Lord

Richard Percy★ (Lecturer in Zoology, Newcastle University); Rt Hon. Enoch Powell, MP; Dr Jane Ridley★ (Lecturer in History at Buckingham University); Professor E. B. Robson★ (Galton Professor of Human Genetics, Department of Genetics and Biometry, University College London); Dr Tom Smith; Rt Hon. Mrs Margaret Thatcher, MP; Dr Martin and Sarah Tolly★; Miss Audrey Tyler★ (Section of Medical Genetics, Welsh National School of Medicine); Members of the United Kingdom Thalassaemia Society; Miss S. E. Walters (General Secretary of the Eugenics Society); Mr Richard Warren★ (Hon. Senior Registrar, Harris Birthright Centre for Fetal Medicine, King's College School of Medicine); Mrs Michael Wheeler-Booth; Mrs Patricia Wilkie★ (Research Fellow and Counsellor at the Polycystic Kidney Unit, Glasgow Infirmary); Professor Bob Williamson (Department of Biochemistry, St Mary's Hospital Medical School, London); Miss Fran Willison (Editor of the *Muscular Dystrophy Journal*).

The mitosis and meiosis diagrams in Chapter 2 appear by kind permission of James S. Thompson and Margaret W. Thompson and W. B. Saunders (Philadelphia); a new edition of *Genetics in Medicine* will be published in 1986. The illustration in Chapter 3 appears by kind permission of Masson S.A. (Paris) and Cambridge University Press (Cambridge); readers may like to know that there is a 1983 edition of Tuchmann-Duplessis's *Embryologie*. George Allen & Unwin (London) granted permission to quote extensively in Chapter 7 from R. Snowden's and G. D. Mitchell's *The Artificial Family* (London, 1981). Ruth Brandon granted permission to quote in Chapter 8 an article which first appeared in *The Observer* on 12 June 1983.

S.F.T., 1985

1 Introduction

*Should we leave the fruits of human reproduction to take shape at
random, keeping our children dependent on the accidents of romance
and genetic endowment, of sexual lottery or what one physician calls
'the meiotic roulette of his parents' chromosomes'? Or should we be
responsible about it, that is, exercise our rational and human choice,
no longer submissively trusting to the blind worship of raw nature?*
(Joseph Fletcher)[1]

Genetic risk is a part of the human condition. Our bodies contain
many thousands of genes. Almost all of these genes consist of long
chains of the chemical deoxyribonucleic acid (DNA), which has
four different chemical subunits. The order in which the subunits are
stringed together spells out each gene's message in the genetic code.
A defect in a single gene can be health-threatening or even fatal.
Some combinations of individually harmless genes lead to handicap.

In Britain, hereditary or partially hereditary disorders account for
perhaps a quarter of all perinatal deaths, three-quarters of serious
handicap and a small but significant proportion of medical problems
outside these categories. Hospital surveys in Canada and in the United
States suggest that this picture is not a rarity. Away from the Third
World the relative importance of environmental as distinct from
genetic factors in mortality and ill health has declined as well nour-
ished, well housed, vaccinated and inoculated populations exert increasingly effective control over nutritional and infectious disease.
And now that the high tide of nutritional and infectious disease has
receded, all kinds of genetic and partially genetic disorders glisten on
the epidemiologist's foreshore – a challenge for medical science and a
challenge for every one of us as parent or prospective parent.

Several thousand genetic disorders have been identified in man.
Some of them cause problems from birth, or even earlier. Others
only become a problem in later life. Many are very minor but the

great majority are serious. Although there are effective therapies for a variety of genetic disorders, most seem to be difficult to treat. And full cures – the repair of specific genetic defects rather than therapy for their effects – are not yet a part of our doctors' armoury.

Some doctors temporise. The prevalence of genetic suffering saddens them. They would like to reduce it in future generations. But they believe that nothing can be done until new means of treatment are developed. Other doctors refuse to wait. They say that facts about the recurrence rates of genetic disorders can be used, and should now be used, as tools for preventive medicine. Many cases of genetic disease would be avoidable if the medical profession did more to educate high-risk couples about their chances of producing affected offspring. At present many couples plan and complete their families without knowing the genetic risks. An early-warning system might reduce the average number of children high-risk couples choose to have, thus reducing the incidence of high-risk conceptions.[2] And in respect of a small but ever-growing number of genetic disorders (these doctors add) there is no longer any need to rely on the deterrent effect of risk figures that are derived from knowledge of a couple's antecedents. Prenatal diagnostic tests can provide knowledge about the conceived but not yet born; diagnosis in the womb and selective abortion of definitely affected or high-risk fetuses should play an increasingly significant part in preventive medicine.

Not every doctor subscribes to one or other of these schools of thought. Some doctors believe it would be morally wrong to make no immediate use of the information that medical genetics can provide. But they are not prepared to regard medical genetics as a branch of preventive medicine. They are convinced that they have a duty to inform individuals and couples of the characteristics, distribution, prognosis, treatment and heredity of certain threatening disorders, and of the option, where it exists, of selective abortion; but the primary goal of this education is a fully informed autonomous couple or a fully informed autonomous individual. Having attempted to convey the relevant information in an unbiased and sympathetic manner, the medical expert must encourage parents and prospective parents to reach their own reproductive decisions. Medical expertise does not entitle a doctor to usurp lay people's responsibility for the creation of life. Couples with, say, a 1 in 4 risk of producing an affected child ought to be aware of this level of risk; but it is for them, and not for their doctor, to decide that such a risk is, or is not, worth taking. The paramount object of counselling is not to modify reproductive behaviour so that couples behave exactly as the counsellor believes he or she would behave in similar circumstances. Counselling is, or ought to be, an impartial exercise. Viewed solely or even primarily as preventive medicine, it all too easily becomes propaganda. And any determinedly 'preventive' counsellor must live with the thought that this method of preventing future cases is strikingly inefficient.

Consider an inherited disorder for which there is no prenatal diagnostic test. If the 'preventive' counsellor claims responsibility for deterring a hundred pregnancies, each of which would have carried a 10 per cent chance of producing a particular incurable disorder, then that counsellor can congratulate himself or herself on having prevented perhaps ten new cases; conversely, he or she must also accept responsibility for the loss of scores of healthy lives. Can results of this order really be called 'preventive medicine'? Doubts about the appropriateness of this term arise even in relation to a disorder which is diagnosable in the first or second trimester of pregnancy.

Diagnostic tests capable of identifying all but a few per cent of affected fetuses give prospective parents and their medical advisers a relatively efficient means of quality control. Willingness to undergo such a test and to abort any fetus that is shown to be affected means that even high-risk couples can end up with numerous unaffected children. Some doctors and their patients extol the virtues of selective abortion; others see it as a morally suspect departure from the traditional life-preserving orientation of medicine. Doctor Leon Kass has noted the philosophical difficulties that arise when selective abortion performed on eugenic grounds is classified as preventive medicine.

> For in the case of what other diseases does preventive
> medicine consist in the elimination of the patient-at-risk?
> Moreover, the very language used to discuss genetic disease leads
> us to the easy but wrong conclusion that the afflicted fetus or

person is rather than has the disease. True, one is partly defined by
his genotype, but only partly. A person is more than his disease.
And yet we slide easily from the language of possession to the
language of identity, from 'He has haemophilia', to 'He is a
haemophiliac', from 'She has diabetes' through 'She is diabetic'
to 'She is a diabetic', from 'The fetus has Down's syndrome' to
'The fetus is a Down's.' [3]

Is the human embryo or fetus a person? [4] The 1984 Warnock
Report on Human Fertilisation and Embryology said that the human
embryo, by or in itself, had no legal status in the United Kingdom. It
was not, under law, accorded the same status as a child or an adult,
and the law did not treat the human embryo as having a right to life.
Several Acts of Parliament gave limited protection in various respects.
Abortion was a criminal offence save in the circumstances laid down
by statute. And damages might be recovered under the Congenital
Disabilities (Civil Liability) Act 1976 where an embryo or fetus has
been injured in the womb through the negligence of some third
person – it is accorded a kind of retrospective status where it is born
damaged or deformed as a result of wrongful injury. [5]

Mindful of recent and potential developments in medicine and
science related to human fertilisation and embryology the Warnock
Committee recommended that the embryo of the human species
should be afforded both statutory recognition and special protection
in law.

The Committee also expressed the hope that Parliament would
legislate so that any unauthorised use of embryos resulting from *in
vitro* fertilisation would in itself constitute a criminal offence. (The
literal meaning of '*in vitro*' fertilisation is fertilisation 'in a glass'; *in
vitro* fertilisation takes place outside the body.) It should be a criminal
offence to handle or to use as a research subject any live human
embryo derived from *in vitro* fertilisation if the embryo had developed for more than two weeks after fertilisation. Up to the end of
the two-week period, lawful research would be possible provided it
was carried out in accordance with conditions imposed by a special
licensing body. In the spring of 1985 the Unborn Children
(Protection) Bill, a private member's measure presented by Enoch
Powell, had its second reading. The Bill went beyond the

Warnock recommendations, seeking to prevent a human embryo
being created, kept or used for any purpose other than enabling a
child to be borne by a particular woman; possession of an embryo
would be an offence unless the possessor had written permission
from the Secretary of State; the Secretary of State would not be
empowered to authorise possession for research purposes. Enoch
Powell's Bill failed to reach the statute book. He predicted, however,
that its lines would be closely followed when the government framed
its own legislation on the subject.

Some parents and potential parents think it absurd that a tiny
embryo should be accorded legal protection. The Warnock recommendations and the title of Mr Powell's Bill put them in mind
of Laurence Sterne's satirical advocacy of rights for homunculi in the
eighteenth century. Homunculi flourished almost exclusively in late
seventeenth century and eighteenth-century imaginations. Impossible to see under even the most advanced microscopes of the day,
they were said to be miniature preformed humans, contained within
male sperm or even (so the ovists claimed) within the female egg.

> The HOMUNCULUS, Sir, in how ever low and ludicrous a light
> he may appear, in this age of levity, to the eye of folly or
> prejudice; – to the eye of reason in scientifick research, he stands
> confess'd – a BEING guarded and circumscribed with rights: –
> The minutest philosophers, who, by the bye, have the most
> enlarged understandings, (their souls being inversely as their
> enquiries) shew us incontestably, That the HOMUNCULUS is
> created by the same hand, – engender'd in the same course of
> nature, – endow'd with the same loco-motive powers and
> faculties with us: – That he consists, as we do, of skin, hair, fat,
> flesh, veins, arteries, ligaments, nerves, cartilages, bones, marrow,
> brains, glands, genitals, humours, and articulations; – is a Being
> of as much activity, – and, in all senses of the word, as much
> and as truly our fellow-creature as my Lord Chancellor of
> England. – He may be benefited, he may be injured, – he may
> obtain redress; – in a word, he has all the claims and rights of
> humanity, which *Tully, Puffendorff*, or the best ethick writers
> allow to arise out of that state and relation.[6]

Yet laughter at Sterne's joke about the status of human gametes does not prevent us from wondering whether embryos and fetuses might be fellow humans. Do such organisms have a right to life? Do they have a right to be killed so as to avoid predictable suffering later on? Are they, so far as the doctors are concerned, truly patients?

A few malformations of the fetus and at least one prenatally detectable error of metabolism can be treated during pregnancy. Fetal therapy means that 'preventive' medicine needn't eliminate the patient at risk. Yet existing and foreseeable opportunities to practise medicine in the womb raise complex issues. Doctors must try to define the benefits and risks of diagnosis and treatment in the womb. For the fetus, the risks of the procedure (which will often include the possibility of accidentally induced preterm delivery) have to be weighed against the possibility of correction or significant amelioration of the malformation. The benefits to be derived from correction depend both on the severity of the malformation and on its predictable consequences on quality of life and survival. Intervention will usually carry some risk for the mother. She may not be willing to undergo a risky operation.

Parents and potential parents who have never given even a moment's thought to genetic disease, or who have formed only a hazy underestimate of the incidence of hereditary and partially hereditary disorders, are sometimes stunned when they first learn that a serious medical problem in the family has genetic implications. This is not surprising. Considered individually rather than in the aggregate, single-gene defects are quite rare, with cystic fibrosis, for example, one of the commonest of the autosomal recessive disorders in Britain, occurring in only about 1 per 2,000 of the population. Huntington's chorea, a late onset autosomal dominant trait affects about 1 in 9,000. Down's syndrome, a chromosome disorder rather than a single-gene disorder, affects no more than about 1 per 650 live-born infants. Patients and their families focus their attention on the single disease or disorder that threatens them, and when they learn that 'their' disease has an incidence of, say, 1 per 2,000, they may use this figure to measure the extent of their misfortune. The initial question 'Why me?' or 'Why our son?' or 'Why our daughter?' can easily become 'Why does our child suffer this disease when 1,999 other children have been spared?'

Academically-minded geneticists and not a few doctors have a reputation for interpreting this sort of question as a plea for a scientific lecture on genes, chromosomes, inheritance patterns and probability. Their response is understandable. In Britain and in North America less than half of all school-leavers have had teaching leading to an examination in biology and, until recently, both elementary and intermediate level biology textbooks paid little attention to human genetics. Even supposedly educated men and women start families knowing little or nothing about genes, chromosomes, cell division and reproductive risk. But when things do go wrong a lecture on biomedical science may answer only some of the needs of the parents and other members of the family.

Unlike many categories of medical disorder, inherited and partially inherited disorders have immense potential for generating feelings of guilt. When parents first realise that the suffering of their child or the unsatisfactory condition of a conceptus was caused by their own genes, their sense of responsibility can be temporarily overwhelming. Many doctors and specialist genetic counsellors try to banish feelings of guilt. They will often stress the randomness in sexual reproduction. A distraught couple may be told: 'The reassortment of genes in procreation is a chance phenomenon and *every* procreative attempt has some element of risk – that's as true for couples who have produced only healthy children as it is for couples like you who have one or several genetically affected children.' This kind of counsel can certainly lighten the burden of personal responsibility. But the psychological cost of shifting the burden may be high.

In cultures in which fatalism is not the norm, one of the most psychologically unacceptable notions confronting individuals at risk of contributing to the transmission of a genetic disease is the thought that they or their children have been or may become passive victims of a totally random genetic accident. Many men and women perceive their lives and the universe as conforming to the laws of cause and effect. True randomness is often unacceptable or incomprehensible on more than an intellectual level. 'In general parlance,' as the American psychologist Wexler has observed, 'we speak of "chance" in the context of being "lucky" or "unlucky" – a personal attribute which, as it were, controls chance and mitigates against randomness. When an individual has been the victim of a violent crime, others often respond with accusations instead of sympathy. The victim is considered to have covertly incited the crime through some careless or inappropriate behaviour. Even the victim often feels ashamed and self-recriminatory. The advantage to this way of thinking is that the crime can thus be attributed to a specific action that can then be avoided by others in the future. If disaster, either natural or man-made, is truly random, then we are all and at all times vulnerable. Some . . . manage their fears by turning to a higher order of control and explanation in the medium of religion. God can be influenced through prayer and good deeds. If He should choose to inflict the disease, then it is not randomly assigned but made meaningful through God's will.' [7]

The Old Testament God made a point of blessing the seed of obedient Abraham and of cursing the disobedient. He was, the Old Testament says, a jealous God, visiting the iniquity of the fathers unto the third and fourth generations. In later writings, medieval Christians such as Boethius and Dante maintain the pagan notion of Fortune alongside a belief in God's omnipotence. And Aquinas taught that the notion of Divine Providence does not exclude the operation of luck. In the fifteenth and sixteenth centuries Protestant theologians ridiculed the idea that the Church could manoeuvre God's grace for earthly purposes. Hardships of earthly life were to be tolerated because of the blessings of the next.

Calvin saw little point in trying to avert plague and thunderbolts through the medium of prayer. They were God's plague; God's thunderbolts. The Elizabethan bishop Thomas Cooper insisted: 'That which we call fortune is nothing but the hand of God, working by causes and for causes that we know not. Chance or fortune are gods devised by man and made by our ignorance of the true, almighty and everlasting God.' [8] Life was not truly a lottery. There was divine concern with every detail. If something went wrong it was meaningless to curse one's luck: God's hand was at work. Believers should respond to personal calamity by searching themselves for moral defects.

Calvin and Bishop Cooper died in the second half of the sixteenth century. The next hundred years or so witnessed the rise of modern experimental science. Nature, hitherto widely assumed to be a sacred seal and witness of God's elementary plan, was found by Galileo to be written in language which any extremely competent, mortal mathematician could understand. By the second half of the seventeenth century Erasmus's 1509 mockery of the sort of person who spends his life 'building countless universes and measuring the sun, moon, stars, and planets by rule of thumb or a bit of string, and producing reasons for thunderbolts, winds, eclipses and other inexplicable phenomena' had begun to seem old-fashioned.[9] Many phenomena were leaving the realm of the inexplicable and were becoming predictable. Nature had fewer secrets. The universe could be seen as an enormous clock, set in motion by God and regulated by him. Europeans became quite obsessed with quantitative phenomena.

Historians in pursuit of the cultural origins of quantified uncertainty seem to be particularly interested in the seventeenth century. Perhaps risk figures only began to inhabit the consciousness of ordinary members of the general public at about this time. Writing in 1662, just three years before the Great Plague, John Graunt ventures a seemingly novel criticism of the people who took in the weekly bills of mortality for the City of London. Graunt complained that these people had been making little other use of these records than to look at the foot; and to note among the casualties what had happened rare and extraordinary. Graunt, no ordinary man, was determined to make better use of the data. 'Whereas many persons live in great fear and apprehension of some of the more formidable, and notorious diseases following; I shall only set down how many died of each: that the respective numbers, being compared with the total 229,250 [the recorded mortality over twenty years], those persons may the better understand the hazard they are in.'[10] This notion that men and women ought not to consider good and evil in itself; but also the probability or likelihood of any hoped for or feared event, appears in the Port Royal *La Logique ou l'Art de penser*, a widely read and extremely influential contemporary work.

The Jansenist authors of *The Art of Thinking* claim that some people panic dreadfully when they hear thunder. Yet if 'It is . . . only the fear of being thunderstruck causes this extraordinary apprehension, then it will easily appear how little reason they have. For of two millions of persons it is very much if one be killed in that manner: and we may also aver, that there is no sort of violent death happens so rarely. Since then the fear of mischief ought to be proportionable to the greatness of the danger, and the probability of the event, as there is no sort of danger that so rarely befalls us as to be killed with thunder, so have we the least reason to fear it: since that fear will no way avail us to avoid it.' [11] The figure of 1 in 2 million represented didactic guesswork, but mushrooming studies of quantified facts were already reducing the areas in which guesswork was necessary. Learned men and women were becoming very interested in the tendency for some experimental and natural arrangements to produce stable long-run frequencies.

John Arbuthnot, physician to Queen Anne and a Fellow of the Royal Society, took the trouble to study the registers of christenings in London. [12] He told his readers: 'It is odds, if a woman is with child, but it shall be a boy; and if you would know the just odds, you must consider the proportion in the bills [registers] that the males bear to the females.' [13] Arbuthnot considered scores of successive years, and found a slight preponderance of male births in every one of them. 'To judge the wisdom of the contrivance, we must observe that the external accidents to which males are subject (who must seek their food with danger) do make a great havock of them, and that this lot exceeds far that of the other sex, occasioned by diseases incident to it, as experience convinces us. To repair that loss, provident nature, by the disposal of its wise creator, brings forth more males than females; and that in almost constant proportion.' [14]

Another Fellow of the Royal Society, William Derham, the rector of Upminster in Essex, was mightily impressed by Arbuthnot's suggestion that a slight average excess of male births over female births had persisted not only in London, but all over the world. 'What,' Derham asked, 'can the maintaining throughout all ages and places these proportions of mankind, and all other creatures, this harmony in the generations of men, be but the work of one that ruleth the world! . . . How is it possible by the bare rules, and blind acts of nature, that there should be any tolerable proportion; for instance, between, males and females, either of mankind, or of any other creature . . .!'[15,16]

In the nineteenth century it was not an English rector but a Moravian monk who sowed the seeds of the modern science of genetics. For much of his life Gregor Mendel was usually dressed in the more or less ordinary clothes that were appropriate for a member of the Augustinian order acting a schoolmaster. He loved to take his pupils round the monastery garden at Brünn. Being sensitive to draughts, he very sensibly hung an Aeolian harp in the garden. Whenever the harp sounded, he put on his hat. Wearing a frockcoat that was too large for him, and short trousers tucked into topboots, Mendel kept in touch with the harp-graced world of Psalm 137. He also explored new territory.

Mendel revolutionised the study of heredity by describing genetic behaviour in probabilistic terms. He assigned Chance a central role in the mechanism of heredity, realising that the by now familiar laws of probability enabled a well-informed person to predict not only the kind of progeny a gardener could expect from a cross between two plants, but also the frequencies with which each kind of plant appeared. Mendel's predictive power was not unbounded. But in respect of a limited number of characters it was really very impressive. Data from extensive breeding experiments enabled the father of modern genetics to see into the future. If you gave him crucial information about an unseen crop's pedigree Mendel would be able to predict that round peas, for instance, would exceed wrinkled peas in the ratio of 3 : 1. Observation of statistically significant samples of peas had taught him that this tended to be the case. The statistical regularity Mendel observed in his breeding experiments, and his new insights into heredity, were set down not in any work of natural theology but in one of the soberest parts of the *Proceedings of the Natural Science Society* of Brünn. From 1866 any bookworm could learn that yet another aspect of life on earth had been explained.

In common with many of the naturalists who were at work in the early and mid-nineteenth century Charles Darwin was quite unaware of Mendel's conceptual breakthrough. Darwin's *Variation of Animals and Plants under Domestication* contains many cases of hybridised breeding which do in fact give 'Mendelian' results. In that work however, as in the more famous *Origin of Species* and *The Descent of Man*, Darwin displays far greater interest in the production of variation than in the laws of inheritance. If Darwin had stumbled across Mendel's hypothesis that the basis of heredity does not blend but is made up of discrete and separable units that recombine in accordance with the laws of probability, would it have been possible for Darwin to die viewing the world as Arbuthnot, Derham and Paley had viewed it? There is no reason to suppose so. The basis of Darwin's belief that 'The old argument [for the existence of God] from design in Nature, as given by Paley, which formerly seemed . . . so conclusive, fails' was Darwin's discovery that 'there seems to be no more design in the variability of organic beings, and in the action of natural selection, than in the course which the wind blows'.[17] The Darwinian and neo-Darwinian world was a restless world, characterised by struggles between members of the same species and by struggles between species. Life was in process. No life form, however finished and perfect it might seem to be, could be regarded as permanent. Chance genetic variation and environmental change made biological change the historical and expected norm.

Mendel's classic report of his own experiments in plant hybridisation went largely unconsidered until the start of the twentieth century. Then it was trumpeted abroad by scientists who had independently performed comparable experiments with similar results, using varieties other than the garden pea. Hugo de Vries, and Carl Correns, early rediscoverers of Mendel's laws of heredity, sought to persuade their fellow scientists of the wide application of Mendel's laws. A significant group was unimpressed. Francis Galton, one of the leading students of human heredity in the nineteenth century, refused to believe that Mendelian-type findings in plants provided a satisfactory basis for a general theory of heredity. Some of the characteristics that Galton had studied were clearly not distributed in the ratios that so excited Mendelist botanists and biologists. It was interesting, certainly, that tall and short pea plants appeared in Mendelian ratios among the offspring of certain crosses, but what had all this to do with humans? Among humans (and many other organisms) stature, for example, is not dished out in the same sharply contrasted way – in a single sibship the heights of adult brothers and sisters may range from, say 6 ft 1 in., through 5 ft 10 in., through 5 ft 8 in., to 5 ft 3 in. According to Galton's Ancestral Law, each parent contributes on average one-quarter to a child's total heritage, whilst the average contribution of each grandparent is one-sixteenth, and so on. The contribution of an exceptionally tall or exceptionally short parent was likely to be diluted by less extreme contributions.

The first years of the twentieth century saw Galton's biometricians struggling for intellectual supremacy over a growing band of convinced Mendelists. William Bateson, a dedicated breeder of plants and one of the earliest English proponents of Mendelism, encouraged doctors to look for Mendelian ratios in families that had medical disorders. In or about 1901, Bateson persuaded Archibald Garrod, who had studied four families in which eleven persons had alkaptonuria (a condition associated with exceptionally dark urine), to consider the hypothesis that there is a recessive heritable factor for this rare human condition. Garrod and Bateson knew that the family trees of patients with alkaptonuria sometimes looked very like the pedigrees that the Mendelists were constructing for plants and animals with recessive features. Garrod said later that the strongest argument that could be adduced in favour of the view that alkaptonuria is a Mendelian recessive character was the fact that albinism, which closely resembles it in its mode of incidence in man, behaves as a recessive character in the experimental breeding of animals. He had in mind the experiments of Castle and Allen, reported in the 1903 *Proceedings of the American Academy of Arts and Sciences*. Garrod used the term 'inborn error of metabolism' to describe an inherited disorder that involved a chemical process.

In 1908 Bateson predicted that 'So soon as it becomes common knowledge – not a philosophical speculation, but a certainty – that liability to a disease, or the power of resisting its attack . . ., is due to the presence or absence of a specific ingredient; and finally that these characteristics are transmitted to the offspring according to definite, predicable rules, then man's views of his own nature, his conceptions of justice, in short his whole outlook on the world, must be profoundly changed'.[18] Bateson seems to have propagated Mendelism everywhere, from YMCA hut to university quadrangle. He was delighted when a Scots soldier summed up one of his popular lectures by saying: 'Sir, what ye're telling us is nothing but Scientific Calvinism.'[19] Bateson wanted everyone to accept that:

> All the ordinary animals and plants began their individual life by the union of two cells, the one male, the other female. Those cells are known as germ-cells or *gametes*, that is to say, 'marrying' cells.
>
> Now obviously the diversity of form which is characteristic of the animal and plant world must be somehow represented in the gametes, since it is they which bring into each organism all that it contains. I am aware that there is interplay between the organism and the circumstances in which it grows up, and that opportunity given may bring out a potentiality which without that opportunity must have lain dormant. But while noting parenthetically that this question of opportunity has an importance, which some day it may be convenient to estimate, the one certain fact is that all the powers, physical and mental, that a living creature possesses were contributed by one or by both of the two germ-cells which united in fertilisation to give it existence. The fact that *two* cells are concerned in the production of all the ordinary forms of life was discovered a long while ago, and has been part of the stock of elementary knowledge of all educated persons for about half a century. The full consequences of this double nature seem nevertheless to have struck nobody before Mendel. Simple though the fact is, I have noticed that to many it is difficult to assimilate as a working idea. We are accustomed to think of a man, a butterfly, or an apple tree as each *one* thing. In order to understand the significance of Mendelism we must get thoroughly familiar with the fact that they are each *two* things, double throughout every part of their composition . . .[20]
>
> To get a true picture of the composition of the individual we have to think how *each* of the two original gametes was provided in the matter of height, hair, colour, mathematical ability, nail-shape, and the other features that go to make the man we know. The contribution of each gamete in each respect has thus to be separately brought to account. If we could make a list of all the ingredients that go to form a man and could set out how he is constituted in respect of each of them, it would not suffice to give one column of values for these ingredients,

but we must rule two columns, one for the ovum and one for
the spermatozoon, which united in fertilisation to form that
man, and in each column we must represent how that gamete
was supplied in respect of each of the ingredients in our list.[21]

[Until Mendel began his analysis] the existence of any orderly
system of descent was never even suspected. In their manifold
complexity human characteristics seemed to follow no obvious
system, and the fact was taken as a fair sample of the working
of heredity.

Misconception was especially brought in by describing descent
in terms of 'blood'. The common speech uses expressions such as
'consanguinity', 'pure-blooded', 'half-blood' and the like, which
call up a misleading picture to the mind. Blood is in some
respects a fluid, and thus it is supposed that this fluid can be
both quantitatively and qualitatively diluted with other bloods,
just as treacle can be diluted with water. Blood in primitive
physiology being the peculiar vehicle of life, at once its essence
and its corporeal abode, these ideas of dilution and compounding
of characters in the commingling of bloods inevitably suggest
that the ingredients of the mixture once combined are
inseparable, that they can be brought together in any relative
amounts, and in short that in heredity we are concerned mainly
with a quantitative problem. Truer notions of genetic
physiology are given by the Hebrew expression 'seed'. If we
speak of man as 'of the blood-royal' we think at once of
plebeian dilution, and we wonder how much of the royal fluid
is likely to be 'in his veins'; but if we say 'he is of the seed of
Abraham' we feel something of the permanence and
indestructibility of that germ which can be divided and scattered
among all nations, but remains recognisable in type and
characteristics after 4,000 years.[22]

If, instead of exhibiting the successive pairs of progenitors
who have contributed to the making of an ultimate individual,
some one [before Mendel] had had the idea of setting out the
posterity of a single ancestor who possessed a marked feature
such as the Habsburg lip, and shewing the transmission of this
feature along some of the descending branches and the
permanent loss of the feature in collaterals, the essential truth
that heredity can be expressed in terms of presence and absence
must have at once become apparent. For the descendant is not,

as he appears in the conventional pedigree, a sort of pool into which each tributary ancestral stream had passed something, but rather a conglomerate of ingredient characters taken from his progenitors in such a way that some ingredients are represented and others are omitted.

I have spoken cautiously as to the evidence for the operation of any simple Mendelian system in the descent of human faculty; yet the certainty that systems which differ from the simpler schemes only in degree of complexity are at work in the distribution of characters among the human population cannot fail to influence our conception of life and of ethics, leading perhaps ultimately to modification of social usage.[23]

In 1918 R. A. Fisher published a paper which took some of the heat out of the controversy between Mendelists such as Bateson and the biometrician students of Galton. Fisher observed that certain seemingly awkward findings of the biometricians in regard to traits such as stature are exactly what one would expect, on statistical grounds, if multiple genes, each having an additive effect and each behaving in a Mendelian way, are involved.

By the end of the First World War, both alkaptonuria and albinism had been interpreted as Mendelian recessive traits on the basis of simple pedigree studies. Doctors also knew that the ABO blood types, discovered at the start of the century, were inherited. The word 'genetics', meaning that part of biological science which is concerned with the study of heredity and variation, had been coined by Bateson, Cambridge University's first Professor of Biology. Many of the terms of present-day medical genetics were in existence.

In the early part of this century the hospitals and medical schools made only a relatively minor contribution to the new science. Scientists knew that they could make the most rapid strides if they concentrated on quick-breeding organisms. The amorous fruit fly *Drosophila* not only breeds quickly; it also has four pairs of very large chromosomes. This made *Drosophila* particularly rewarding to study since it was realised that chromosomes, structures found inside cells, were likely to be the transporters of Mendel's factors, or genes. The flies had two further advantages over humans. Controlled breeding meant that there was little risk of patterns of inheritance being obscured either by unrecognised cases of non-paternity or by multiple deaths from malnutrition and infection. *Drosophila* became the geneticists' favourite pet. From about 1910, Thomas Hunt Morgan and co-researchers at Columbia University in America established that factors were located on the chromosomes of *Drosophila*. Intensive breeding experiments indicated that *Drosophila*'s genes were arranged in a characteristic linear sequence along the length of its chromosomes. Genes located near together on a chromosome tended to be inherited, if they were inherited at all, along with their close neighbours. Morgan received a 1933 Nobel Prize for this work.

The first compulsory course in genetics for American medical students seems to have been instituted at about this time. In 1936 J. B. S. Haldane and Dr Julia Bell started investigating how close was the link between the human gene carrying colour-blindness and the human gene causing haemophilia. Genetics did not become firmly established as part of the British medical syllabus until about 1938, however. Research on human chromosomes was hampered by bench-level difficulties and by the relatively slow rate of human breeding. Human genetics would remain in the scientific doldrums until the new science of molecular biology and new methods for studying human chromosomes excited researchers in the 1950s. In the early 1960s the number of papers listed in a major monthly bibliography of the literature of biomedicine under such headings as human heredity, chromosomes and human genetics exceeded the number devoted to hernias. Medical genetics left the doldrums, set on a course which has revolutionised both medicine and parental choice.

Improved methods for studying human chromosomes led to the demonstration in 1956 that the usual human complement of chromosomes is forty-six in each body cell. A few years later, workers in France and Britain showed that Down's syndrome and a number of other clinically recognisable abnormalities were the product of certain easily detectable chromosome aberrations. This had immediate implications for prenatal diagnosis.

With the discovery in 1944 that purified deoxyribonucleic acid (DNA) molecules were capable of transmitting hereditary features from one bacterial cell to another, the concept of the gene had undergone a dramatic change. No longer was it merely an element of formal genetic analysis. Genes were palpable, perturbable entities and there was no reason why scientists should not try to unravel their structures. Watson's and Crick's 1953 proposal for the structure of DNA revealed how DNA could replicate and encode important information in the linear sequence of its nucleotides.

Pauling and others discovered that the inherited disease sickle-cell anaemia is due to an abnormal haemoglobin molecule: in 1956 their hypothesis that the relevant change was due to an amino acid substitution was confirmed. Many inborn errors of metabolism were found to owe their origin to enzyme deficiencies. Scientists identified the particular enzyme defect in one form of phenylketonuria. In this case the new understanding led to the development of effective preventive therapy. Generally, however, the picture was less bright. It was proving difficult to treat successfully many of the disorders that were known to be wholly or partially genetic.

In the 1970s direct analysis of the gene became practicable through the application of recombinant DNA technology. Whereas classical Mendelian genetics had dealt mainly with the inheritance of easily observable characteristics the new genetics could deal with the chemistry of the gene itself.[24] This opened new doors to carrier detection and to prenatal diagnosis. In the much longer term the new genetics might also lead to the repair of specific gene defects.

2 Genes, Chromosomes and Patterns of Inheritance

The linkage between human generations is cellular.
(Clifford Grobstein) [1]

Genes and chromosomes

Like the bodies of elephants and the bodies of mice, human bodies are made up from millions and millions of cells. Virtually every body cell contains a complete set of hereditary factors called *genes*. Although these factors (long chains of deoxyribonucleic acid (D N A)) are chemically inert, their function is vital. They determine the synthesis and structure of *proteins*, which are the machine tools, building materials and engines of living cells. If we are to understand how the cells in our fingers accomplish the feat of being genetically matched with the cells in our toes, we must remember that the linkage between generations is cellular.

We started life as a *zygote*, a solitary cell no larger than a pinpoint, which was formed when a sperm fused with an egg.[2] Within the centre, or *nucleus*, of the zygote there lay a microscopically visible network of dark-staining material called *chromatin*. It was this material, carrying our genes, that was chiefly responsible for directing our embryonic and later development along human rather than elephantine or mouselike lines. Each time a cell division occurred, first splitting the zygote into two cells which themselves split (making four cells) which themselves split (making eight cells) which themselves split (and so on) the chromatin and our genes took great care not to be left behind. As the zygote prepared to divide, the chromatin arranged itself into forty-six *chromosomes* and allowed each chromosome to split longitudinally. Consequently when the zygote cleaved into two cells, two sets of the original genetic instructions were available – one set for each cell. Repeated *mitosis* made it feasible for virtually all our body cells to have a complete set of genes.

If the copying of genetic material is perfect at every cell division, our body cells will be equipped with forty-six chromosomes arranged in twenty-three pairs. Although the chromosomes in each pair are similar they are not identical. The non-identity of each pair is explained by the fact that each parent has contributed a chromosome to it. A zygote derives twenty-three chromosomes from the mother's egg (one representative of each chromosome pair) and twenty-three chromosomes from the father's sperm (one representative of each chromosome pair).

The formation of egg cells and sperm cells (our *germ* cells) is governed by a special procedure called *meiosis*. In meiosis the total number of chromosomes passed to each cell is half the total found in the ordinary body, or *somatic*, cell. Human germ cells therefore have just twenty-three chromosomes, one from each pair. Chance alone determines which member of a pair of chromosomes arrives in any given germ cell. There are about 8 million different possible chromosome combinations in the germ cells of a single individual. The number of possible combinations of chromosomes in a zygote of any given set of biological parents is about $8,388,608 \times 8,388,608$; and the potential genetic variety is further enhanced by *crossing-over*, the exchange of genetic material during pairing of homologous chromosomes in meiosis. This means that the chromosomes of an individual's germ cells are different from those of either parent.

> Every child conceived by a given couple is the result of a genetic lottery. He is merely one out of a large crowd of possible children, any one of whom might have been conceived on the same occasion if another of the millions of sperm cells emitted by the father had happened to fertilise the egg cell of the mother – an egg cell which is itself one among many. And all of these possible children would be as different from one another as the actual ones.[3]

The sex of our children is determined by the sex chromosome in

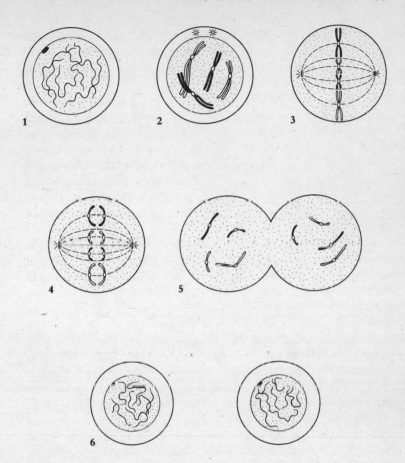

Mitosis

Simplified diagram showing replication of two of our twenty-three chromosome pairs. Although the chromosomes in each pair are similar, they are not identical. Each parent has contributed a chromosome to the pair. In this diagram chromosomes from one parent are shown in black; chromosomes from the other parent appear in outline. (After James S. Thompson and Margaret W. Thompson, *Genetics in Medicine*, Philadelphia, 1980.)

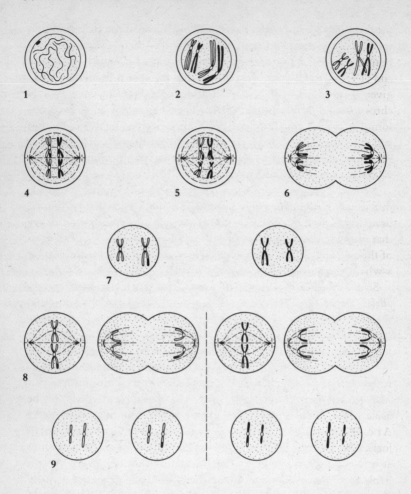

Meiosis

Simplified diagram showing meiotic division of two of our twenty-three
chromosome pairs. Crossing-over, the exchange of genetic material during pairing
of homologous chromosomes, may occur at stage 3. (After James S. Thompson
and Margaret W. Thompson, *Genetics in Medicine*, Philadelphia, 1980.)

the fertilising sperm. Males have an X and Y chromosome in addition to their forty-four non-sex chromosomes in each somatic cell. X-bearing and Y-bearing sperms are manufactured in similar quantities and there is roughly an even chance that the sperm involved in any given fertilisation will be a Y-bearing one. Females have two X chromosomes in their somatic cells. Their eggs carry an X chromosome. A zygote which inherits an X chromosome from the mother and a Y chromosome from the father will develop as a boy. Girls are produced when the father has added a second X chromosome to the X chromosome the mother provides.

Normal inherited differences between individuals are due to the differences in the genes they possess. If genes are likened to minute beads strung upon lengths of thread (the chromosomes) we can say that normal gene differences usually involve extremely small sections of thread. The only notable exception to this rule is the difference of a whole chromosome between the sexes.

Scientists refer to alternative forms of a particular gene as *gene alleles*. At every position, or *gene locus*, on our paired chromosomes we may have identical alleles or different ones. In any large population there will be many different alleles for each locus but let us suppose that the reader belongs to a population in which there are just two alleles, A1 and A2, for a given locus on a chromosome pair. If the same allele is carried on both members of the chromosome pair (the reader carries A1 A1 or A2 A2) the reader is said to be *homozygous*. If the reader carries different alleles (in this example, A2 A1 or A1 A2) he or she is *heterozygous*. There are tens of thousands of loci on our twenty-three pairs of chromosomes. We are homozygous at some and we are heterozygous at others. Although every allele is liable to change, or *mutate*, to another form, our genes are actually remarkably stable. At some loci the observable mutation rate is only about ten changes per million genes per generation.[4]

Benign genetic variations such as the A, B, AB and O blood groups, rarely concern us. Readers with blood type 'O' happen to be homozygous, having inherited two 'O' genes, one from each parent. By contrast, readers with the AB blood group are heterozygous, having acquired an 'A' gene from one of their parents and a 'B' gene from the other. Most of us never have cause to think about the character and origins of the alleles at these particular loci – in everyday life, one blood group is just as good as another. But many loci have alleles which are not benign. About 1,500 human disorders are known to be caused by single mutant genes. For the members of the unfortunate families that are affected by the more serious of these *monogenic* disorders the distinction between the homozygous and the heterozygote state may be too important to be ignored.

Autosomal recessive inheritance

Let us suppose that a quite serious disorder is diagnosed in the 11-month-old daughter of Mr and Mrs Brown. The Browns fear that the daughter's brother, who is 6 years old and healthy, will catch the condition if the children continue to live in the same house and play together. The doctors allay this fear by explaining that the daughter's problem is neither infectious nor contagious. It is a genetic disorder of infancy, they say, and the brother's six years of rude health means that he will never be troubled by it. Although Mr and Mrs Brown are grateful for this news (the thought of separating the children appalled them) they are also puzzled. Their daughter's condition is genetic. That's clear enough. But their first child, their son, is as much a product of their genes as their daughter is, so how has he managed to escape his sister's fate? If the Browns have another child, as they are planning to do, will that child have the same disorder as the daughter has, or will it escape?

If Miss Brown's condition is one of the hundreds which follow an *autosomal recessive* pattern of inheritance her parents' wonderment at her brother's good health need not last long. Recessively inherited disorders occur when an individual receives not just one potentially harmful gene at a particular locus, but two. People who carry only one potentially harmful gene at the locus may well go through all or much of their life without being aware of their carrier status. The basic mechanism in many autosomal recessive diseases is enzyme deficiency; heterozygotes often have about 50 per cent of normal enzyme activity. The mutant gene's power to harm them is kept within tolerable bounds by the healthy influence of the normal matching gene on the other member of the chromosome pair. In the case of Miss Brown the condition is known to be one which follows an autosomal recessive pattern. We may therefore infer that she has a potentially harmful recessive gene at the same locus on each member of one of her twenty-two pairs of non-sex chromosomes (*autosomes*). As she received half her genes and chromosomes from her mother and the other half from her father it would seem that both Mr and Mrs Brown carry the mutant gene.[5] Their son's good health is explained by the fact that he inherited at least one healthy matching gene. The diagram (page 38) makes the position clear and shows that the third child Mr and Mrs Brown are planning will have a 3 in 4 chance of avoiding the combination of genes which gave rise to the daughter's condition.

When the time comes for Miss Brown to think about having children of her own she will know that her homozygous state means that it is likely that every one of her egg cells carries a copy of the potentially harmful gene. If she marries a man with two healthy genes at the locus all her offspring will be heterozygous carriers of the harmful gene but none will inherit the disorder. If she were to marry a carrier, however, each of her children would have a 1 in 2 chance of inheriting a second copy of the harmful gene and any who did would show symptoms in due course. Offspring of a union with a fellow sufferer would receive two copies of the harmful gene, one from each parent, and all would be affected.

The medical prospects of the children of the brother of Miss Brown will be determined by his actual genotype. We have seen that he either has two normal genes at the locus or a normal gene matched with and masking an abnormal one. If he has two normal genes, every sperm cell he manufactures should carry a normal copy and there is therefore no substantial risk that his children will have the disorder. Marriage to a non-carrier would obviously be problem-free; marriage to a carrier would simply mean that each child had a 1 in 2 chance of being a carrier; and marriage to a sufferer would mean no more than that every child would be a carrier. Suppose instead that the brother actually carries one abnormal gene and one normal one. What are the implications of this genotype? Well, marriage to a non-carrier would not involve any substantial risk of having an affected child – each child would simply have a 1 in 2 chance of

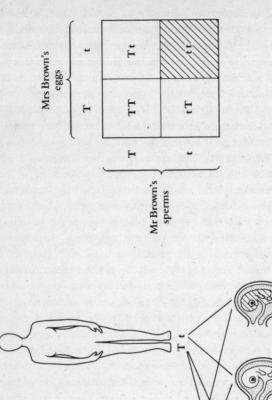

T – normal gene
t – recessive mutant gene

being a carrier; marriage to a fellow carrier would place the brother and his spouse in exactly the same position in respect of risks as his parents, Mr and Mrs Brown, were in – every child would have a 1 in 4 chance of inheriting the disorder. Carrier status and marriage to a woman with not one but two abnormal genes at the same locus would give every child of the marriage a 1 in 2 chance of having the disorder.

When medical geneticists contemplate the long list of human recessive disorders and the frequency of a small but significant group of these disorders they are driven to the conclusion that most of us must carry at least one potentially harmful recessive gene. Indeed the average number of such genes may be four or five (at different loci). Far from being freaks, Mr and Mrs Brown were simply carriers like the rest of us. Their daughter's misfortune stemmed from the fact that both parents carried a potentially harmful gene at the *same* locus. Even then there was still a 3 in 4 chance that any given zygote would be spared the genetic constitution which gave rise to her disorder.

In the case of some of the most intensively studied recessive disorders, it is possible to arrive at quite well-informed estimates of the statistical odds against a Mr and Mrs Brown type of marriage. We know, for instance, that the recessive gene for thalassaemia major is carried by about 1 in 6 people in Cyprus; by 1 in 14 people in Greece; by 1 in 30 people in India and by 1 in 1,500 native Britons. In the United States the recessive gene for sickle-cell anaemia is carried by about 1 in 12 blacks. About 1 in 25 Americans of northern European origin carry the recessive cystic fibrosis gene; and about 1 in 30 American Ashkenazi Jews carry the recessive gene for Tay-Sachs' disease. On the basis of these and other figures it has been estimated that when an American black marries another, unrelated, American black there's a 1 in 150 chance that they both carry the sickle-cell disease gene. When one American of northern European origin marries another, unrelated, American of northern European origin there's a 1 in 620 chance that they both have the gene for cystic fibrosis. In non-consanguineous marriages between American Ashkenazi Jews there's a 1 in 900 chance that both spouses have the gene for Tay-Sachs' disease.

People in Mr and Mrs and Miss Brown's position sometimes rail against their own ancestry, overlooking the fact that descent from

Autosomal recessive pattern of inheritance

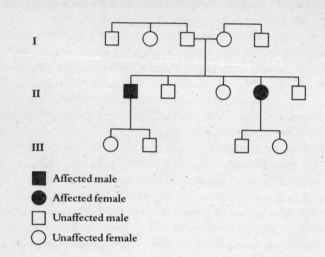

different stock would not have made procreation risk-free. It would merely have reduced some recessive risks and increased others.

Autosomal dominant inheritance

The genes that are responsible for the next group of disorders we must look at are rarer than the potentially troublesome autosomal recessive genes. Autosomal *dominant* as distinct from recessive disorders can trouble individuals who have but one harmful gene at some locus on one of the twenty-two pairs of autosomes. The influence of a harmful autosomal dominant gene is sufficient to offset completely, or to an extent which is not easily tolerated by the body function or structure involved, the healthy influence of the normal matching gene on the other member of the autosome pair. Since all these extra powerful, or dominant, genes are quite rare if the disorder they cause is serious, and since no major ethnic population is saturated with them, it is quite unusual for two carriers of the same harmful autosomal dominant gene to meet, mate and procreate. If there is a previous family history of an autosomal dominant disorder the suffering is likely to have been on just one side of the family.

Let us suppose that a Mr Green is known to carry a harmful autosomal dominant gene – he has grown up with one of the least debilitating of the autosomal dominant disorders, dentinogenesis imperfecta. His wife, Mrs Green, has completely normal teeth and there is no reason to suppose that she carries the dentinogenesis imperfecta gene. When their first child, a boy, is born, it is found that he has dentinogenesis imperfecta and his parents want to know whether any future child will have the same condition.

The diagram on page 42 shows that it makes no difference if Mrs Green transmits one rather than the other of her two healthy genes. Each gene is fated to be dominated if a fertilising sperm carries a copy of Mr Green's harmful gene and each gene will produce a normal dental result if it is paired with Mr Green's normal allele. In every future pregnancy there will be a 50–50 chance that Mr Green contributes a sperm carrying the dentinogenesis gene and consequently every future child (boys and girls alike) can be said to have a 1 in 2 chance of inheriting the disorder. When Mr Green's son and any other child who may inherit the gene reach adulthood and think about starting families of their own, they too will have an even chance of passing a copy of the harmful gene to their offspring. By contrast, unaffected siblings who have not inherited the gene will be no more likely to have affected children than members of the general population.

Exactly the same principles would apply if it was Mrs Green who carried and was affected by the dentinogenesis gene. Like autosomal recessive genes (the sort carried by Mr and Mrs Brown) autosomal dominant genes are just as likely to be transmitted by females as they are by males. To make this point absolutely clear Mr and Mrs Green are featured in the diagram on page 43. This time *Mrs* Green appears as the affected parent.

About a thousand different autosomal dominant diseases have been found in humans. Some of these have far-reaching effects; others are relatively harmless. A few of the severer ones make it unlikely or even impossible for the patient to have children and where this is the

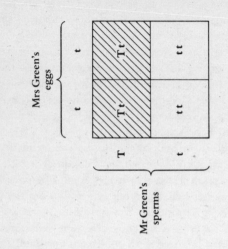

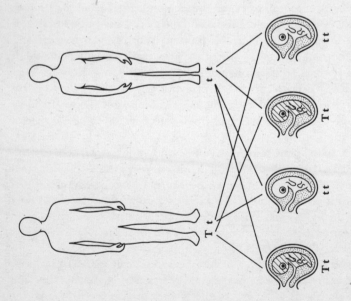

t – normal gene
T – dominant mutant gene

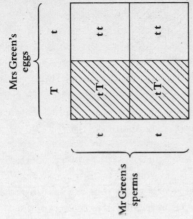

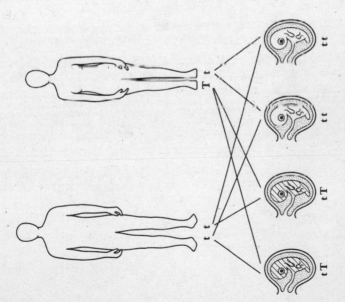

t – normal gene
T – dominant mutant gene

case many of the patients will be new mutants, having received their copy of the relevant gene as a *fresh mutation* in one of the sperms or eggs of a genetically normal parent. Fresh mutations are not peculiar to autosomal disorders. Biologists with an interest in the ways in which helpful as well as unhelpful traits arise in man and other species attach enormous importance to apparently accidental alterations in genetic material. Some non-scientists think 'mutation' a sinister word, but there is no biological reason why the person with a freshly minted version of a particular autosomal dominant gene should feel more miserable, when looking to the future, than the person who has inherited a copy from an affected parent. In both cases the risk that the affected person will produce affected offspring is identical: 1 in 2 in each pregnancy (assuming as before that the mate does not have the same rare gene). In the rare instances where two patients with the same dominant disorder meet and mate there is a 1 in 4 chance of a child homozygous for the abnormal gene. These are usually very severely affected.

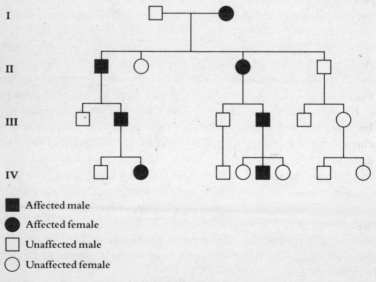

■ Affected male

● Affected female

☐ Unaffected male

○ Unaffected female

Autosomal dominant pattern of inheritance

X-linked inheritance

When harmful or potentially harmful genes are carried on one of the two sex chromosomes rather than on the autosomes the patterns of inheritance are markedly different from the autosomal patterns considered above. It will be recalled that females have a pair of X chromosomes in each somatic cell whereas males have an X and a Y chromosome. On the printed page both sex chromosomes are conventionally represented by standard capital letters. In our cells, however, there is a very significant difference in their respective lengths. The X chromosome is actually different from the Y chromosome in structure as well as length. Many genes on the X chromosome have no corresponding genes on the Y chromosome.

Although both males and females are capable of being affected by X-linked disorders, the sex ratio of affected members in any family will depend on the sex of the parent carrying the gene. In females troublesome genes which are carried on the X chromosome (X-linked) usually act as recessives: in order to express an X-linked disorder a female generally needs to be homozygous for the troublesome gene.[6] In males the absence of a second X chromosome means that there is no normal gene to partner and offset a troublesome X-linked one: hemizygous men and boys are affected. (As males have only one X chromosome they are said to be hemizygous rather than heterozygous or homozygous if they carry an X-linked gene.)

An affected father cannot transmit his X-linked genetic disorder to his sons. The sex chromosome he passes to them is always the Y chromosome. His daughters will have a copy of his troublesome gene, since they receive a copy of his X chromosome. Provided that their mother does not contribute a second copy of the same gene the daughters will usually be unaffected heterozygotes. If any daughter marries an unaffected man and has children of her own, each child will have a 1 in 2 chance of receiving a copy of the troublesome gene. In other words the daughter's daughters have a 1 in 2 chance of being carriers. The daughter's sons have a 1 in 2 chance of being affected.

Since females usually have to be homozygous for a troublesome X-linked gene before it is possible for them to be affected they are *much less commonly* affected than males. On the rare occasions when the homozygous state occurs in a female the genetic prospects for any offspring will depend on the offspring's sex and their father's genetic constitution. If the father is healthy, which means (leaving aside the possibility of a fresh germinal mutation) that he has no copy of the troublesome gene, all the daughters of his affected wife will be healthy heterozygotes and all the sons will be affected. By contrast if the father does have the gene and the disorder, all the daughters and all the sons will be affected.

A simple diagram cannot encompass every situation in X-linked inheritance. Only two situations will be examined in detail. One shows the genetic implications when a single X-linked gene is carried by a mother; the other brings into focus the implications when a father carries an X-linked gene.

In the diagram on page 47, a Mrs Gray has a potentially troublesome haemophilia gene on one of her two X chromosomes. As her husband is healthy he can be depended upon to have a copy of the normal matching gene on his X chromosome.[7] Any daughter has a 1 in 2 chance of inheriting a copy of Mrs Gray's haemophilia gene. If she does, she will be an unaffected carrier. Each son has the same 1 in 2 chance of inheriting a copy. As this will not be partnered by a gene from the father's X chromosome (sons receive the father's Y chromosome) every son who receives a copy will be affected.

In the diagram on page 48 it is not Mrs Gray but Mr Gray who has the X-linked haemophilia gene. He is affected by it but the outlook for all the children is much better than it was in the first illustration. No son will receive a copy of Mr Gray's X chromosome and thus none will inherit the gene and the disorder. All daughters will receive a copy of Mr Gray's X chromosome and the troublesome gene which is on it but in every case this gene will be partnered by one of Mrs Gray's two healthy genes. Daughters will therefore be unaffected carriers.

Queen Victoria of England carried an X-linked haemophilia gene. Albert, the Prince Consort, was unaffected. Each of their five daughters had an initial 1 in 2 chance of inheriting a copy of the gene and it is thought that two of the daughters – Princess Alice and Princess Beatrice – received a copy.[8] Neither one of these princesses

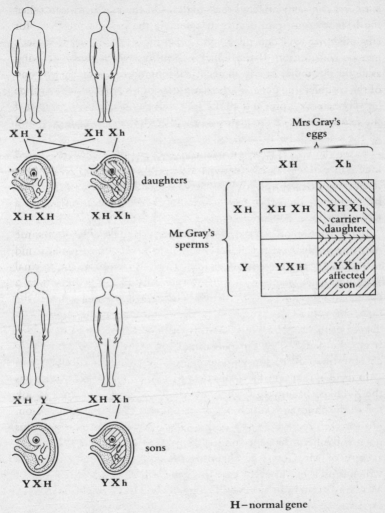

Mrs Gray's eggs

daughters

Mr Gray's sperms

sons

H – normal gene
h – recessive mutant gene

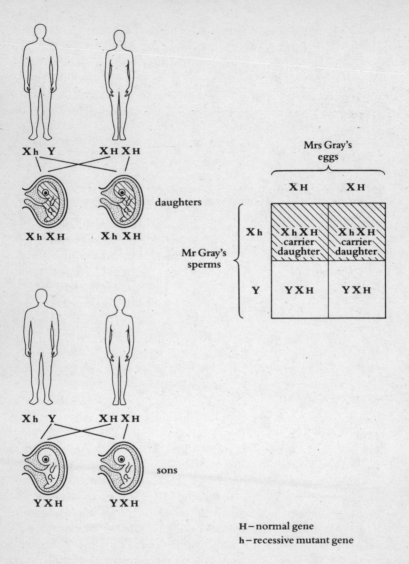

daughters

Mrs Gray's
eggs

sons

		X H	X H
Mr Gray's sperms	X h	X h X H carrier daughter	X h X H carrier daughter
	Y	Y X H	Y X H

H – normal gene
h – recessive mutant gene

had haemophilia. Each of their four brothers had an initial 1 in 2 chance of inheriting the mutant gene. Leopold actually did so. He suffered from haemophilia and transmitted the haemophilia gene to his daughter's son. The haemophiliac Tsarevitch Alexis of Russia received his copy of the X-linked gene from the Tsarina Alix who was the daughter of Princess Alice and the granddaughter of Victoria.

More than a hundred different X-linked disorders and traits have been found in humans. The genes involved are rarer than deleterious autosomal recessive genes and reports of fresh mutations are far commoner in X-linked inheritance. Modern students of the pedigrees of Queen Victoria and Prince Albert are virtually certain that the royal cases were the result of a fresh mutation. But at the time when Leopold's haemophilia was diagnosed the science of clinical genetics had yet to be invented. Medical men knew that haemophilia ran in families but they had no exact knowledge of the incidence of the disease in the general population and they may well have underestimated the frequency of what the modern geneticist means by the term 'fresh mutation'. After Victoria was told of the hereditary nature of Leopold's condition 'a cloud of worry and bewilderment ... overhung the Queen, caused by her oft-repeated and perfectly correct belief that haemophilia was "not in our family" – meaning the House of Hanover. Where did it come from?'[9]

Y-linked Inheritance

Scientists have worked out the inheritance pattern of a disorder associated with a gene on the Y chromosome. Assuming that such a gene had no normal counterpart on the X chromosome, an affected father would be expected to transmit the disorder to none of his daughters but to all of his sons. In turn none of his sons' daughters would be affected but all of his sons' sons would be. To date no such gene for a serious disorder has actually been traced to the Y chromosome and the pattern is therefore of theoretical rather than practical significance.

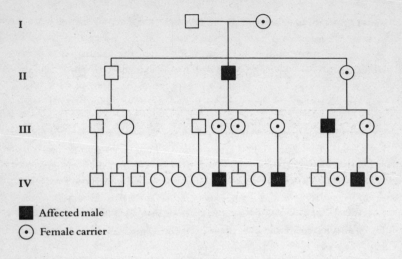

■ Affected male

⊙ Female carrier

X-linked recessive pattern of inheritance (males reproducing)

Chromosomal disorders

In all of the patterns of inheritance considered so far – the X-linked, (the Y-linked), autosomal dominant and autosomal recessive – everything has hinged on genes or pairs of genes at a single locus. No less 'genetic', but usually having a very much lower risk of recurrence in any given family, are the large number of *chromosomal disorders*. Whenever these disorders appear whole clusters of genes are implicated. In Down's syndrome, which is probably the best-known of all the chromosomal disorders, patients have a complete extra chromosome (number 21) in all or many of their cells. Genes on the triplicated chromosome 21 cannot work in pairs. Instead, some sort of triple alliance is formed and the clinical features bear witness to the difficulties.

The great majority of children with chromosomal disorders are born to parents who do not themselves suffer from any of the medical problems which have come to be associated with certain chromosome

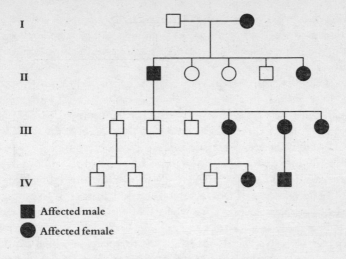

■ Affected male

● Affected female

X-linked dominant pattern of inheritance (not discussed in the text)

abnormalities and it is clear that things sometimes go wrong during the formation of the parental egg cells and sperm cells. Normal germ-cell formation (*gametogenesis*) requires the chromosomes in the first or reduction division (*meiosis*) to separate to opposite sides of the dividing cells. When this does not happen and, for example, both members of a pair of chromosomes go to the same side, one of the two new eggs (or new sperm cells) will receive an extra chromosome. Eggs and sperms with such abnormalities probably have a smaller chance of being involved in zygote formation than normal eggs or sperms but when they do meet up with a normal spermatozoon or egg and become fertilised, a *trisomic* cell (one with an extra chromosome) or a *monosomic* cell (one with a chromosome missing) will be formed. Not every trisomy and monosomy arises in this way. Some arise during early cell division of a normal zygote.

Cells with a chromosome missing tend to be less viable than cells which have an extra chromosome. Monosomies in live births are rarer than trisomies. The large majority of embryos with too many or too few chromosomes undergo spontaneous abortion, usually

during the first trimester of pregnancy and sometimes at such an
early stage that the woman involved is unaware that there has been a
fertilisation, let alone a natural weeding out of a chromosomally
aberrant embryo.

It is unusual for an infant to be born with monosomy for an
autosome but there are many cases of viable infants with just a single
X chromosome. Trisomies involving the autosomes are not uncommon but sex chromosome trisomies (XXX, XXY or XYY),
which tend to be less debilitating, are more numerous.

For some reason non-disjunction of a chromosome during
germ cell formation is particularly associated with high maternal
age.[10] The incidence of Down's syndrome in babies born to mothers
aged 25 is under 1 in 1,000 live births. The figures for mothers aged
38 and for those aged 44 are 5 in 1,000 live births and 25 in 1,000 live
births respectively. The risk does not rise to a level in which amniocentesis would usually be considered until a woman is in the second
half of her thirties.

In addition to monosomies and trisomies there are chromosomal
disorders which have structural origins. From time to time a
chromosome breaks in two places and is carelessly repaired by nature.
The two ends are reattached when the segment is still upside down.
When this happens important genes can find themselves placed out
of sequence and when they take their cue from their new neighbours
the usual miracle of perfect heart or head formation in the uterus
may be prejudiced.

It is not unknown for two different chromosomes to break and for
the fragments to exchange positions so that, say, chromosome
number 21 has a piece of chromosome number 14 grafted on and
number 14 incorporates a bit of number 21. When such exchanges
involve the loss or gain of a segment of chromosome the result is an
unbalanced translocation and serious birth defects are likely to result.
Balanced translocations, as translocations involving no gain or loss of
chromosome material are called, very often cause so little disruption
to the bodies of the people who carry them that their presence goes
unsuspected until the birth of a defective baby prompts an investigation. Depending on which chromosome is involved a prospective father with a balanced translocation may have about a 3 per cent

risk of producing abnormal offspring while the risk in the case of a prospective mother with a balanced translocation is around 10 per cent. In the rare situation where a prospective parent has two 21 chromosomes translocated on to each other the risk of abnormal offspring is 100 per cent in every pregnancy.

Some of us have two quite distinct populations of cells in our bodies – cells with normal chromosomes and cells with chromosome aberrations. Such chromosomal mosaicism accounts for some of the exceptionally mild cases of a recognised syndrome, and is caused by non-disjunction during cell division of the early embryo. When the chromosomally normal cells are present in all but a few of the patient's organs and tissues, the deleterious influence of the chromosomally abnormal cells will usually be patchy.

Only a small proportion of all the chromosomal disorders that obstetricians, paediatricians and other medical professionals encounter are due to parental chromosomal inversions and translocations. And many couples who seek genetic counselling after the birth of a child with a chromosomal abnormality are relieved to be told that a fresh non-disjunction or chromosome rearrangement is responsible: 'It's just been an accident, not something wrong with our genes so it isn't really a genetic thing in our family.'[11]

Multifactorial disorders[12]

Chromosome aberrations and single mutant genes account for rather less than half of the birth defects which medical professionals see each year. The majority of birth defects seem to be determined by a combination of genetic and environmental factors. Disorders which have a partly genetic and partly environmental cause do not exhibit such clear-cut patterns of inheritance as the autosomal recessive, autosomal dominant and X-linked disorders. They do tend to cluster in families but as more than one genetic locus is involved and as even a genetically predisposed individual may avoid exposure to the largely unidentified environmental triggers, the risks of recurrence in affected families are often significantly lower than the risks in autosomal dominant, X-linked and autosomal recessive inheritance.

When considering multifactorial inheritance it may be helpful to think in terms of large numbers of extremely common genes, each having an additive effect. In every family, as we have seen, zygotes normally draw half their complement of chromosomes from the egg and half from the fertilising sperm. *On the average*, first-degree relatives have half their genes in common, while second-degree relatives (for example aunts, nieces and nephews) and third-degree relatives (for example first cousins) have respectively a quarter and an eighth of their genes in common. This helps to explain why the incidence of multifactorial disorders is higher in first-degree relatives of patients than in more remote relatives.

Before we look at several specific risk estimates it is essential to appreciate that the incidence and therefore the risk of recurrence of any given multifactorial disorder may vary from one population to another, for both genetic and environmental reasons. Quite recently, multifactorially caused defects of the neural tube (which include spina bifida) occurred in the United Kingdom at a rate of around 5 per 1,000 births. But in Ireland, Scotland and Wales the incidence was about 7 per 1,000 whereas in London the frequency was less than 3 per 1,000. In the United States the incidence was thought to be around 2 per 1,000 births overall, with lower frequencies for orientals and blacks. Not every group of multifactorially caused disorders varies in frequency to this extent but frequency variations as between one population and another are exceedingly common and the best estimate of a recurrence risk is likely to be one that takes account of recently gathered local data.

In autosomal recessive, autosomal dominant and X-linked inheritance, once it has been established that a given individual or couple is genetically equipped to produce offspring with specific troublesome or potentially troublesome genotypes, the probability of succeeding children having these genotypes is constant. For instance, if we know that Queen Victoria carried a haemophilia gene on one of her two X-chromosomes, we know also that Leopold, the eighth of her nine children, had exactly the same chance of inheriting a copy of the haemophilia gene as the others. Each fertilisation was an independent event, with two equally likely outcomes: either the queen would transmit a copy of her normal X-linked gene or she would transmit a copy of the haemophilia version. Looking back we can appreciate that Edward VII's good fortune, Alice's misfortune and the good fortune of Alfred and Arthur had no influence whatsoever on the odds for Leopold and his younger sister Beatrice. Each child had the same 1 in 2 chance of inheriting a copy of the haemophilia gene. This constancy of probabilities among siblings threatened by certain monogenic disorders contrasts strikingly with the recurrence risks in multifactorial inheritance. We shall see that the frequency of multifactorial disorders in subsequent children does appear to rise, quite significantly, after successive affected children have been born.

Let us suppose that Mr and Mrs Scarlet belong to a population in which the incidence of cleft lip (with or without cleft palate) is 1 per 1,000 births, with girls and boys being affected with much the same frequency. Mr and Mrs Scarlet want to know whether the three children they are hoping to have would be likely to arrive with cleft lip (with or without cleft palate). Assuming that neither of the Scarlets is affected in this way and assuming also that there are no affected close relatives, the best estimate that can be offered *before* the first child is born may be the general population risk of 1 in 1,000. If the first child turns out to be unaffected, the same estimate will be appropriate for the second planned child. However if the first child does have cleft lip (with or without cleft palate), the risk estimate in respect of the second child will have to be changed.

Although the predisposing genes in question are thought to be extremely common, with perhaps the majority of the population having quite a sprinkling of them, only a minority of matings actually yield affected offspring. Therefore if the Scarlets were to have an affected first-born, demonstrating an abundance of predisposing genes between them, it would be sensible, perhaps, to classify their marriage as one which is a little exceptional in genetic terms and to estimate the risk for the second child on the basis of the incidence of cleft lip (with or without cleft palate) among children who have an affected brother or sister. Scrutiny of available data might show that, on the average, 1 in every 33 such children were affected. This would give the Scarlets' second child a risk some thirty times higher than the normal population risk of 1 per 1,000. If the Scarlets were unfortunate enough to have two affected children with cleft lip (with or without cleft palate) then the risk for the third child would be estimated by studying other couples with two affected children and if the incidence among children with two affected siblings was say 1 in 11 (again on the average) the risk estimate for the Scarlets' third planned child would be about a hundred times higher than the normal population risk.

These values may be quite close to the values in the reader's own population but they are offered as illustrations and should not be used as the sole basis for important personal or professional decisions. When estimating occurrence and recurrence risks for particular families, the skilled geneticist takes account of a host of additional factors such as the severity of the disorder in relevant patients, the clinical condition of parents and even, on occasion, the clinical condition of close relatives. Some multifactorial disorders are known to affect boys more commonly than they affect girls. With others girls are the more commonly affected.

In the United Kingdom congenital pyloric stenosis, a multifactorially caused disorder of the gastrointestinal tract, is about five times as common in boys as in girls. The boys, it seems, need fewer predisposing genes to trigger environmental factors than the girls. Put another way, the girls require a greater genetic liability to develop pyloric stenosis than do the boys. On the occasions when a girl is affected it is probably reasonable to assume that the parents have a quite exceptionally large supply of predisposing genes between them. Subsequent children of the marriage will therefore be exceptionally likely to inherit a high dose of the relevant genes. The effect of such a dose depends to some extent on the sex of the individual concerned. Girls can tolerate higher predisposing doses than boys, so if the second child is a girl the risk that she will be affected will be lower than the risk for a male second-born (about 4 and 9 per cent respectively).

With pyloric stenosis the risks following the birth of an affected first-born son are lower than the risks after the birth of an affected first-born daughter. Because it doesn't need such a large dose to trigger the environmental factors for boys a single male index case is perfectly consistent with the hypothesis that the parents have an above average total of predisposing genes between them but not a really heavy concentration of them. In certain populations the recurrence risk for the second-born will be about 3 per cent if the second-born is a girl and around 4 per cent if a boy. These risk estimates reflect the incidences in a number of published studies but there are many variables, and no reader should assume that these estimates would be appropriate for their own family.

Warning

Readers who are interested in their own genetic risks or in the risks of a potential mate or in the risks of close family members would be well advised to consult a professional geneticist. Without professional advice it is all too easy for the amateur student of genetics to over-estimate or underestimate a risk.

Experienced clinical geneticists are very careful to avoid the trap of unsuspected *genetic heterogeneity*. They know that certain disorders show autosomal recessive inheritance in some families and autosomal dominant inheritance in others – the clinical condition of an individual is not always an indication of his genotype. Geneticists also know that variability is common in genetic traits. This means that two individuals with the same mutant gene are not predestined to suffer in exactly the same way. The severity of a genetic disorder often varies considerably, even within a single family.

In some situations it will be quite impossible for even the most expert clinical geneticist to be certain about the mode of inheritance in a particular case. If, for instance, Mr and Mrs Mystery have a child with a serious disorder and scientists know, by consulting the relevant literature, that this disorder sometimes occurs more than once in a family, although in the majority of cases there is only one affected family member, then the Mysterys' geneticist has a knotty problem. The geneticist will first try to establish whether there are any other similarly affected individuals in the family. Suppose, however, that inquiries covering living relatives and past generations reveal no other affected individual. In this situation the geneticist has no way of knowing whether the ill health of the Mysterys' first child is an isolated one-off case, or if there is a significant risk for subsequent children. If Mr and Mrs Mystery are demanding to know the risk of producing a second affected child the geneticist may consider the advantages and disadvantages of at least four different answers.

Mr and Mrs Mystery could be told that it is impossible to know the risk of recurrence. Alternatively, their geneticist might be prepared to advise that the risk is virtually 0 per cent if the first child was affected out of the blue but 25 per cent if Mr and Mrs Mystery are both carriers of an autosomal recessive gene. A third approach would be to tell Mr and Mrs Mystery that statistics on all families with a child with this disorder reveal an empiric recurrence risk of about 5 per cent. On the other hand, the geneticist could avoid any mention of specific figures but indicate that the risk of recurrence was 'relatively low'. Readers may like to think about these options. Which answer would be the most helpful? Why? [13]

Geneticists sometimes turn to a simple blood test or to a test on easily available tissues to establish whether a key person is in fact a carrier of an important gene. Thousands of Jewish adults have been screened voluntarily to see if they have the gene for Tay-Sachs' disease. A small percentage were shown to carry the gene. Screening has also been instituted for thalassaemia major and sickle-cell anaemia. And it seems likely to be widely available for cystic fibrosis in the quite near future. Not every test achieves 100 per cent accuracy in clinical practice but an increasing number of tests are sufficiently accurate to be used in genetic counselling. With their assistance many, though by no means all, of the clinical geneticist's uncertainties can be removed.

The list of disorders for which carrier detection is feasible will grow. Parents and potential parents with an interest in the subject should make sure that their knowledge is up to date.

3 Conception and Development

Of Bodies, and of Mans Soul.   (*Sir Kenelm Digby*) [1]

Fertilisation of an egg by a sperm, the yoking together of maternal and paternal genes, usually takes place in one or other of the two fallopian tubes. The zygote that is formed travels along the tube towards the uterus, wrapped in a protective acellular coat called the zona pellucida. The zygote divides into two blastomeres at about 30 hours after fertilisation. [2] The cells continue to divide and form a ball, or morula, of sixteen or more cells. The morula enters the uterus. Fluid from the uterine cavity finds its way into the morula and at about 96 hours after fertilisation the cells rearrange themselves to form the blastocyst.

At the blastocyst stage the fertilised egg resembles a hollow ball. The cells at the periphery (trophoblast cells) will contribute to the placenta. A knob of cells (the inner cell mass) attached to the inner side of the trophoblast will give rise mainly to the embryo. The blastocyst lies free in the womb for about two days. When it loses its protective acellular wrapping, some 96–120 hours after fertilisation, it is ready to implant itself in the womb lining (endometrium). During the process of implantation, which spans six to seven days, trophoblast cells invade the endometrium. Endoderm forms at about 168 hours after fertilisation from the cells of the inner cell mass which are nearest to the blastocyst cavity. Fluid-filled spaces appear in the remaining inner cell mass and these spaces coalesce to form the amniotic cavity. Between the two cystic spaces within the blastocyst,

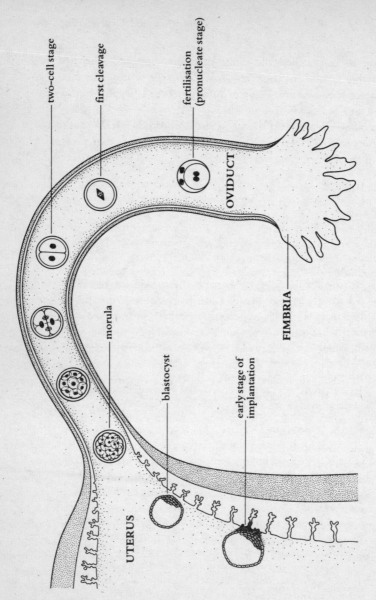

two-cell stage

first cleavage

fertilisation
(pronucleate stage)

OVIDUCT

FIMBRIA

morula

blastocyst

early stage of
implantation

UTERUS

Development of the human embryo

(After H. Tuchmann-Duplessis, G. David and P. Haegel, *Illustrated Human Embryology*, Vol. 1, New York/London/Paris, 1971.)

a plate of cells called the embryonic disc appears. The first recognisable features of the embryo will emerge within this disc.

One of these features is the primitive streak, a heaping-up of cells
at one end of the embryonic disc.

An eloquent pregnant diarist whose embryology is vivid, though
not always orthodox, describes some of the changes that take place
during the first twelve weeks or so: [3]

> By the ninth day after conception the single cell which is the
> egg has become multi-celled and already differentiated into the
> three prime body layers from which will come all the organs,
> different systems from each layer. [4] From the outermost layer,
> the ectoderm, will come skin, hair, eyes, nails, nervous system;
> from the mesoderm, the circulatory and supporting, or muscular
> and skeletal systems; from the inner layer, or endoderm, the
> digestive apparatus, including liver, lungs, bladder, stomach and
> thyroid. [5] What marvellous organisation! Not quite so quick as
> God's creation of the universe, but at least everything knows
> what it is going to be by the ninth day. And, moreover, this
> differentiation is basically alike in all vertebrate animals. Fish,
> turtle, fowl, hare, and human; all are similar thus far. The
> variations come later. [6]
>
> By eighteen days, the heart, nervous system, and head end
> have been marked out and are recognisable. Somewhere
> between the third and fourth weeks, probably, the heart begins
> to beat. Toward the end of the fourth week the embryo has
> grown so much that it has to start to curl up – the position in
> which it will remain until it is born. At this time it has a tail,
> gill clefts, and the beginnings of eyes and a nose. During the
> fifth week arm and leg buds appear. At six weeks the five chief
> divisions of the brain appear. [7]
>
> Up until this time the embryo is still very small and can rest
> easily in a teaspoon, membranes and all. But at about four weeks
> the maternal–fetal blood exchange becomes well established, and
> the fetus begins to get more nourishment. Thereafter it begins
> to grow much faster, and by six weeks it is about the size of a
> silver dollar, and its presence can be detected by a physical
> examination for the first time. [8]
>
> But by now, more than three months old, it is growing
> rapidly – a millimetre and a half a day. According to my

embryology book, a child of ten who had continued to grow at this rate, small though it may seem to be, would be twenty feet tall. And its face is definitely human, even having eyelashes and eyebrows. Its external genitalia have developed, too, so if I could only see it I would know immediately what sex it is. And it has fingers and toes and ears.

Obstetricians often call the first twelve weeks or so of intra-uterine development the 'first trimester'. At the start of the second trimester (which stretches from about week 13 to about week 24) rapid growth is one of the most notable features. Growth slows during weeks 17–20. In this period fetal movements known as quickening may be felt by the mother for the first time. The fetus puts on weight in weeks 21–25. All organs become quite well developed, but the respiratory system is still immature.

Shortly after the start of the third trimester the lungs become capable of breathing air and the central nervous system acquires the ability to control body temperature. If a fetus is born after only 26–28 weeks' gestation its viability is jeopardised to some extent by the incomplete development of its respiratory and central nervous systems. Even so its chances in an artificial incubator and respirator are now very much better than they have been.

By full term, about 38 weeks after fertilisation, the fetus weighs about 3 kg.

Many scientists agree that the first nine months of development must involve about forty-two cell divisions. There is, however, no scientific consensus on the point at which a human person comes into existence. Religious leaders cannot agree. Pragmatic legislators bypass the issue. Even spouses argue. The beginning of a person is not, apparently, a simple question of fact. Scientific observation, no less than theological and philosophical speculation, throws shadows.

From ancient times right up until the second half of the nineteenth century the basic cellular facts of human reproduction were unknown. Aristotle in common with every other biologist of the pre-microscope era had no opportunity to detect microscopically small spermatozoa in male seminal fluid. Human ova are tiny; far smaller than the eggs of birds and fishes and barely visible to the naked eye. Seeking an analogy to the bird or fish egg in humans and other mammals in which he had seen no egg, Aristotle compared the embryo in its membranes to an egg. Some of his followers believed that organisms formed through sexual reproduction received their substance from the female 'egg' and their form from the male semen: the semen was thought to bring about a coagulum of menstrual blood before getting to work like an artist, imparting human shape and form to the raw materials provided by the female. Lucretius (*c.* 60 B C) took a different view: 'Births always consist of two-fold seed; and, whichsoever of their parents children more resemble, they of that one have a more than equal share.'[9] Galen (A D 129–199) and many early Arabian thinkers were quite sure that there must be a material contribution by the male.

Lacking any idea of cell division – the process which enables zygotes to become multi-celled organisms within days of their formation – many Aristotelians took the view that the fetus does not become ensouled or distinctly human until the moment of quickening. This view did not go unchallenged. Platonists, for instance, believed that the soul entered at conception and the Stoics, in common with some Hindus, believed that it did not arrive until the moment of birth.

A number of early Catholic thinkers gravitated to the Aristotelian middle ground, and for much of the Church's history the commoner opinion among the most learned churchmen may have been that the soul delays its arrival until some time after conception. Augustine (354–430) maintained a distinction between the ensouled and the unensouled fetus but said he did not know exactly when ensoulment occurred. Aquinas (1225–74) followed Aristotle's schedule and declared that the rational soul enters the male fetus at about the fortieth day after conception; female fetuses were not ready for their souls until very much later. Some theologians were careful to distinguish man's rational soul from more lowly forms. Living, growing and feeding plants had mere vegetative souls; they lacked consciousness. Organisms that were conscious but not self-conscious possessed animal souls. Only humans were self-aware. Their distinctive rational soul was implanted by God when the embryo was well enough developed to receive it.

Semen was thought to be blood without impurities. Dante's *The Divine Comedy* reflects this view:

The perfect blood, which never is drunk up
    Into the thirsty veins, and which remaineth
    Like food that from the table thou removest,
Takes in the heart for all the human members
    Virtue informative, as being that
    Which to be changed to them goes through the veins.
Again digest, descends it where 'tis better
    Silent to be than say; and then drops thence
    Upon another's blood in natural vase.
There one together with the other mingles,
    One to be passive meant, the other active
    By reason of the perfect place it springs from;
And being conjoined, begins to operate,
    Coagulating first, then vivifying
    What for its matter it had made consistent.
The active virtue, being made a soul
    As of a plant (in so far different,
    This on the way is, that arrived already),
Then works so much, that now it moves and feels
    Like a sea-fungus, and then undertakes
    To organise the powers whose seed it is.
Now, Son, dilates and now distends itself
    The virtue from the generator's heart,
    Where nature is intent on all the members.
But how from animal it man becomes
    Thou dost not see as yet, this is a point
    Which made a wiser man than thou once err
So far, that in his doctrine separate
    He made the soul from possible intellect,
    For he no organ saw by this assumed.
Open thy breast unto the truth that's coming,
    And know that, just as soon as in the foetus
    The articulation of the brain is perfect,
The primal Motor turns to it well pleased
    At so great art of nature, and inspires
    A spirit new with virtue all replete,
Which what it finds there active doth attract
    Into its substance, and becomes one soul,
    Which lives, and feels, and on itself revolves.
And that thou less may wonder at my word,

> Behold the sun's heat, which becometh wine,
> Joined to the juice that from the vine distills.[10]

Theories of delayed ensoulment were still current when Fienus, a professor of medicine at Louvain, published a more or less systematic biomedical work on the formation of the fetus. Writing in 1620 Fienus could not accept the received idea that the rational soul arrives only forty or more days after conception. He knew, or thought he knew, that semen coagulates the menstrual blood in just three days. Some sort of a soul had to be present on the third day, to direct the embryo's development. Fienus insisted that this must be the rational soul. He was not dismayed by his inability to adduce any evidence of rational function in the three-day-old conceptus. He believed that adherents to the forty day theory suffered a no less grave difficulty as there was no evident rational function at forty or eighty (or even ninety) days.[11] Indeed, Fienus went so far as to say that rational function is not discernible until some two or three years after birth.

When the philosopher Locke essayed a definition of a person in the same century, he too stressed rationality. For him a person is 'a thinking, intelligent being, that has reason and reflection, and can consider itself, the same thinking thing, in different times and places'.[12]

Today, if a Vatican Declaration issued in the autumn of 1974 is followed, the Roman Catholic Church teaches that 'respect for human life is called for from the time that the process of generation begins. From the time that the ovum is fertilised a life is begun which is neither that of the father nor that of the mother; it is rather the life of a new human being with its own growth.'[13] The belief that female fetuses experience animation and ensoulment much later than male fetuses seems to have gone out of fashion. The Holy See has declared that the question of when there comes into existence an immortal soul is a philosophical problem from which the Church's moral affirmation about the right to life of the human being from conception is independent:

> Even supposing a belated animation, there is still nothing less
> than a *human* life (as biological science makes evident), preparing
> for and calling for a soul for the completion of the nature

received from the parents; on the other hand, it suffices that the presence of this soul be *probable* (and the contrary will never be established) in order that the taking of life involves accepting the risk of killing a human being who is not only waiting for but already in possession of his soul.[14]

Biological science suggests that the life-expectancy of the average fertilised egg is really quite short – an estimated 10–18 per cent of fertilised eggs are lost during the first week of their existence, 12–33 per cent die in the early post-implantation stage and 9–13 per cent are aborted spontaneously as recognised miscarriages. If we believe that every one of these fertilised eggs is a person, arguably possessed of a soul, then we live in a world in which a third to two-thirds of mankind never sees the light of day.[15]

Human twins are of two kinds. Some arise from different eggs. Others, monozygotic twins, originate from a single fertilised egg. Does the possibility of monozygotic twins make it unreasonable to say that every fertilised egg is a new human being? A cellular division occurring as late as twelve days after the fusion of the male and female pronuclei in the centre of an ovum can divide the inner cell mass of the blastocyst into two embryonic primordia. This means that irreversible biological individuality is not assured until about two weeks have elapsed since fertilisation.

Once implantation has occurred, biological individuality is less problematical. The organism's long-term survival is not guaranteed, but any egg which implants successfully should be congratulated for having survived eleven to thirteen days of considerable natural peril. By six weeks after fertilisation, the probability that a particular surviving embryo will give rise to a live infant has risen significantly: about eight out of ten will succeed. By eighteen weeks the success rate is about 97 per cent.

Embryologists sometimes call the embryo a fetus when eight weeks of gestation counted from the first day of the woman's last menstrual period have elapsed (this is usually equivalent to six weeks after fertilisation). Only then, according to some of the textbooks, does something that is recognisable as a human come into being. The transition from a bunch of cells which have potentials to an organism that is visibly human is not abrupt, and a convention of this kind may be rather a shaky basis for decisions about the cells' moral status. Readers may like to test their own responses to images of six-weeks', seven-weeks', nine-weeks', eleven-weeks' and twelve-weeks' development. Photographs can be found in almost any modern embryology textbook. At what prenatal stage, if any, do the cells evoke in the reader empathy as another self?

Early neural development began, as the pregnant diarist noted, in about the third week after fertilisation. The neural groove appears by about the seventeenth day; by the twenty-second to twenty-third day the neural folds can be discerned, and these soon fuse to form the recognisable beginnings of the spinal cord. When does the embryo or fetus become capable of feeling pain? Scientists have noted responsiveness to stimuli as early as seven to eight weeks after conception. At about seven weeks an embryo's nervous system and neck muscles may be sufficiently developed for it to be able to detect slight pressure around the mouth, and to take slow and weak avoiding action, turning the head away from the stimulus. At about eight weeks the response is stronger. Later on, local rather than total-pattern reflexes can be elicited, involving the limbs and other parts of the body. Stereotypic reflex responses are evidence of the spinal cord's growing neurological maturity, but they do not demonstrate higher brain function.

Ultrasound studies have detected apparently spontaneous movements in twelve-week human fetuses. The purpose of such activity is obscure. Young chick embryos exhibit movements which do not appear to be responses to external stimuli. Such behaviour must assist muscular development. Does it imply some degree of sentience? In humans it is common for the mother to feel both spontaneous and reflex fetal movements after sixteen weeks of development. 'Tonight, lying in bed, I realise that what I thought was gas was lower down, more definite – a small, uncontrolled, irritated motion. Oh, I know so well what it is doing now! There was a film I saw once of a very young cat embryo; its movements were like the snap of an uncoiling shrimp, a motion involving the whole body. The separate movement of the limbs comes only much later, after the nervous system is better developed. But even this faint but definite proof of life is rather awesome.'[16]

Throughout the latter half of the pregnancy, human fetuses spend a third or more of each day making breathing movements. These movements coincide with periods of rapid eye movement.

The usual intensity of sound inside the human womb is thought to be approximately 50 decibels (very much quieter than a disco — about the same as a quiet office). Maternal heartbeats are capable of producing 95 decibels. Thermal stimuli are generally absent or quite minor, with the temperature inside the womb remaining at about 0.5 °C above the mother's temperature. The protective cushion of amniotic fluid and the state of near weightlessness reduce the opportunities for tactile stimuli. And the normal intensity of light is similar to that in a darkened lecture room. How dull! Perhaps brain development is influenced to a greater degree by chemicals reaching it from the maternal circulation than from the meagre information about the immediate environment that reaches the brain through the nervous system.

The higher mental faculties are correlated with the development of the cerebral cortex which contains millions and millions of cells and which continues to develop in the post-embryonic stage. Our power to learn from experience, our ability to store and reprocess facts, our consciousness and our creativity arose long after the moment female and male pronuclei came together in the centre of a fertilised egg.

4 Prenatal Tests

*The once seemingly science fiction idea of routine, detailed fetal
analysis in all pregnancies now is upon us. (Thaddeus E. Kelly)* [1]

Until recently, midwives and doctors knew very little about the
developing embryo and fetus. Chaucer's doctor of physic apparently
relied on the standard medieval Latin work on obstetrics *Trotula*,
which may have been written by a female doctor in the eleventh
century.[2] English translations of this work appear in manuscripts of
the late fourteenth and early fifteenth centuries. Surviving late-
medieval illustrations of life inside the womb tend to be hardly less
imaginative than the numerous bestiaries of the period: fetuses often
have the plump calves of a two-year-old, the arm muscles of a
successful pugilist or some other feature that makes them seem old
before their time.

Leonardo da Vinci, perhaps the best anatomist in the late fifteenth
and early sixteenth centuries, resorted to dissection in his attempt to
'describe the nature of the womb, and how the child inhabits it, and
in what stage it dwells there, and the manner of its quickening and
feeding, and its growth, and what interval there is between one stage
of growth and another'.[3] His copious notes and illustrations leave us
to guess how many gravid human uteruses he actually saw. Pregnant
ruminants clearly inspired some of his pictorial accounts of early
human development. He learned also by examining hens' eggs.

Some of Leonardo's drawings of a human fetus in the uterus
fetched up in the Royal Collection at Windsor in the latter part of
the seventeenth century; and a large number of Leonardo's manu-

scripts on both human and comparative anatomy reached this country quite soon after the artist's death. Nevertheless a bookish midwife in Stuart England was more likely to be guided by *The Birth of Mankinde*, one of the earliest printed English textbooks for midwives, which appeared in numerous editions from the 1540s until about the time of the Great Fire of London. Paré's French work on obstetrics was translated into English in the 1630s. William Harvey's work on obstetrics, *De partu*, appeared in an English translation in the 1650s. An altogether less specialised work, *Aristotle's Compleat and Experienc'd Midwife*, was published at the turn of the century and ran to at least ten editions. This popular English text owed rather less to Aristotle than its title suggests.

Embryology certainly attracted a great deal of research and scientific speculation in Tudor and Stuart England. But firm, uncontroversial advance was hampered by several factors. Effective microscopes were not in common use until quite late on. Postmortem studies on English women who had died either early in pregnancy or in childbirth seem to have been rather uncommon. Admitting defeat, some early-eighteenth-century authors of books for midwives drop embryology altogether from their pages.

In more recent years important data began to be gathered from numerous studies on spontaneously aborted embryos and on women who died during pregnancy. But the living embryo or fetus was still pretty unapproachable. It could be listened to. Leonardo had assumed that the fetus's heart does not beat inside the womb. Nineteenth-century doctors were equipped with new-fangled monaural and binaural stethoscopes which made it easy for the sharp of hearing to monitor fetal heartbeats in the later stages of pregnancy.[4] And as more and more was known about the anatomy of the female cervix, the laying on of hands by an experienced midwife or doctor had an increased potential for supplying useful information about the state of the pregnancy.

In the twentieth century X-rays came into use as a novel diagnostic and research tool. The results of fetal and maternal exposure to high levels of radiation were not understood at first and radiography was restricted when the hazards of some of the earliest practices became known.

Amniocentesis, the pricking of the fluid-filled membranous sac that surrounds the fetus in the womb, so that some of the fluid can be examined, enabled doctors from about the middle of the twentieth century to diagnose certain genetic disorders prenatally. In the 1950s it was demonstrated that the sex of a fetus can be determined by staining the nuclei of some of the fetal cells that float in the amniotic fluid. The ABO blood group of a single fetus could also be determined using cells from the fluid. Sexing of the fetus is more reliable if the living cells that are withdrawn are cultured in a special fluid and allowed to multiply for a period of ten to fourteen days. There should then be enough cells for a variety of chromosome studies. Additional culturing can be made to produce enough cells for biochemical studies.

Amniocentesis is rarely attempted before the second trimester of pregnancy. A fetus takes fourteen weeks or so to outgrow the restricted pelvic area and to move up, inside its amniotic sac, to the roomier abdomen. Only when the sac is up in the abdomen does drawing off 5–25 ml of the fetus's 175–225 ml of amniotic fluid become safe enough and simple enough to be regarded as a standard medical procedure. The placement of a hypodermic needle into the womb near the pregnant woman's navel usually involves only minor physical discomfort, although pain is not unknown. Some practitioners and their patients prefer not to use even a local anaesthetic. Large-scale studies in Canada, the United States and the United Kingdom suggest that the procedure carries minimal physical risks for the mother. There is, however, a risk of inducing a spontaneous abortion, and this is the outcome in about 1 per cent of operations.

Some men and women believe that amniocentesis must be a major operation, requiring a general anaesthetic and elaborately equipped operating rooms and a long stay in hospital. Couples may even forgo the procedure because of unwarranted fears about what is involved. It is not unusual for women who have direct experience of amniocentesis to express astonishment at the speed, simplicity and painlessness of the procedure: 'I felt a couple of tiny pin-pricks and thought they were just the preliminaries ... Then I felt a niggly sensation deep inside. It suddenly dawned on me that this was the actual amniocentesis procedure itself.'[5] Psychological reactions are quite variable. One woman who did not return for a second amniocentesis appointment after the first operation had failed to produce an assayable specimen explained that the first attempt at prenatal diagnosis had made her feel tremendously protective towards her child: 'The needle in my stomach felt so threatening.'[6]

About 90 per cent of all amnioceteses are performed for cytogenetic analysis. This means that once the required amount of amniotic fluid has been drawn off, the woman who has undergone the procedure must wait, commonly for two to three weeks, for the results. This is the length of time laboratory technicians need if they are to bring the growth of fetal cells in the amniotic fluid to the stage at which chromosomal analysis is practicable. Sometimes the cells refuse to grow and multiply in the culture fluid. This is not, in itself, an indication of fetal abnormality. It is merely an inconvenient reminder that the successful culture and analysis of cells requires skill from the technicians, excellent laboratory conditions and a small amount of luck. Whenever the first sample fails to produce usable cells, the pregnant woman and her advisers will have to consider the advantages and disadvantages of a second attempt. The later that amniocentesis is performed, the more likely it is that the fetus will have quickened before the results of diagnostic chromosomal or biochemical studies come through. Quickening does not invariably excite awe, but it has been known to constitute a psychological watershed. Once a woman has felt the fetus move inside her, selective abortion may seem less impersonal.

When the results of amniocentesis do come through, 100 per cent accuracy cannot be guaranteed. Scientists have to allow for the possibility of human error in the laboratory; for the possibility that the maternal rather than the fetal cells have been analysed; and for the inherent limitations of the relevant analytical techniques. Even so, a growing number of disease-specific tests are achieving impressive accuracy in clinical practice and news of a favourable result may allow a couple who have been burdened by their knowledge of a high risk to relax for the first time in several months. Knowing at last that the tested pregnancy is almost certainly free from a threatening inherited metabolic disorder, chromosomal disorder or open neural tube defect, they may be able to enjoy the whole of the third trimester and what remains of the second trimester. They will probably be advised that a good result in respect of the particular disorder or group of disorders they were concerned about does not rule out the possibility that the fetus has some other undetected and undetectable disorder, but since this degree of uncertainty is not peculiar to them, but is present in every human pregnancy, they will be told that they are in exactly the same position as other prospective parents are in.

When a couple is worried about a gene defect which cannot be detected in amniotic fluid or cultured cells but which can be detected in fetal blood, ultrasound may be used to locate the placenta to facilitate placental aspiration. Alternatively, aspiration may be done under direct vision, using a fibre-optic instrument inserted into the womb. Fetoscopy also enables doctors to visualise quite a large number of anatomical malformations affecting the limbs, face and genitals. In this context the diagnostic power of fetoscopy rarely arises before the eighteenth week of pregnancy and the advantages of its utilisation have to be weighed against a 5 per cent or so risk of accidentally inducing an abortion.

The pitch of ultrasound is outside the range of human hearing. During ultrasonography the pregnant woman lies on her back and a metal arm is moved painlessly and silently over her oiled abdomen. Waves or pulses of high-pitched sound are then concentrated and emitted, like a beam of light, into the maternal and fetal tissues. The resulting echoes are mapped out, on a cathode ray tube, giving a result like a slightly fuzzy black and white television picture. On first acquaintance, the displayed image of the cross-sectional anatomy can be rather disappointing. 'For all I know I could have been looking at a shoulder of lamb,' said one woman, reporting her first ultrasound glimpse of her fetus.[7] But with greater experience these strange, static cross-sectioned images can be exciting. Two-dimensional real-time scanning, a more advanced form of ultrasonography, enables lay people and professionals alike to make immediate sense of a moving, hitherto private world. If extensive ultrasonography is proved to be absolutely safe and if an ultrasound scan becomes a routine part of every pregnancy it is quite possible that the psychological watershed of felt quickening will be rivalled by a new, sometimes earlier and no less significant watershed: 'It was fascinating to see the tiny arms and legs pounding away vigorously even before I could feel any movements. I couldn't have faced an abortion after that; it would have been a terrible decision.'[8]

Ultrasound enables doctors to recognise a small number of major structural malformations prenatally (anencephaly for instance) but its primary medical uses are: locating the placenta and the fetus's head during amniocentesis (so that these areas are avoided by the needle); identifying multiple pregnancies; and establishing an accurate gestational age and recognising fetal death. With rare exceptions each fetus in a multiple pregnancy has its own amniotic sac. It is sometimes necessary to sample the amniotic fluid in each sac.

The maternal and fetal risks of the existing invasive diagnostic procedures and the limited amount of information that can be supplied by any one technique make doctors reluctant to use invasive investigations on a random basis. At present, some of the major indications for prenatal tests are: advanced maternal age (about 37 years in the United Kingdom and about 35 years in the United States); the existence of one or more children with a neural tube defect; certain chromosomal translocations and other structural abnormalities in either parent; severe detectable autosomal or X-linked recessive metabolic disorders; and severe X-linked recessive disorders which are not detectable but which can be prevented from appearing in the next generation if the parents are willing to abort male fetuses.

Prenatal diagnosis would obviously be transformed if doctors could obtain informative fetal cells early in the first trimester, without the risks that attend amniocentesis. Scientists know that sloughed cells of fetal origin enter the mother's blood circulation after implantation. Some researchers believe that it may be possible to develop a technique to separate these fetal cells from maternal blood and to use the isolated cells to establish fetal cell lines for diagnostic analysis. If such a technique proved successful, then a simple blood sample from the mother would be all that was needed to get the diagnostic procedure started.[9] Another first trimester approach might be via fetal cells washed from the endocervical canal of pregnant women.

Safer techniques for obtaining fetal blood will not only allow doctors to detect more widely disorders in which a molecular defect is expressed in the blood components, but such techniques will also allow analysis of plasma proteins and red cell markers. And once the locus for a particular disorder is shown to be closely linked to the locus for a genetic marker such as a blood group, it will be possible in certain families to predict whether an individual with the marker has inherited the gene for the disorder.

Prediction by genetic linkage analysis is potentially applicable in an informative family to disorders with unknown biochemical defects, and to disorders whose underlying defect is not expressed in cultured cells. Development of DNA probes which are specific for regions of chromosomal DNA in and around genes will increase the number of potential 'markers' of genetic disorders: since recombinant DNA techniques require only minute amounts of tissue, accurate prenatal diagnosis may be possible early in the first trimester, using new techniques of fetal cell sampling.

Chorionic villus sampling (CVS) can already be practised in the eighth to twelfth weeks. This method of prenatal diagnosis calls for a very narrow suction tube to be passed through the vagina into the womb, so that several of the chorionic villi – projections of tissue at the placenta – can be collected for analysis. Ultrasound pictures facilitate the operation. Cells in this tissue usually contain exactly the same arrangement of genetic material as the fetus itself has in each of its body cells. By extracting and analysing DNA from villus tissue, it is possible to establish fetal sex and to detect defects in the fetus's genes. Since the cells are drawn from the membranes there is no need to puncture the amniotic sac. Anaesthesia is not necessary. One of the centres which have pioneered this technique is the Harris Birthright Centre, King's College Hospital, London. Other centres are beginning to undertake it. Maternal and fetal risks are still being evaluated. In some cases a transabdominal approach is indicated.

Chorionic villus sampling screening for Down's syndrome can be performed from about eight weeks (ten weeks is the optimal time) – it is a relatively simple matter to obtain visible chromosomes from the cells for counting. Detecting detailed abnormalities hidden inside the chromosomes is less straightforward and more time-consuming but it can be done. DNA in the cells can be cut up using enzymes. The fragments can then be spread across a gel plate. The DNA is then transferred on to a sheet and it is incubated in a bag of liquid containing pieces of DNA which can hunt for the defective gene and pair with it. The hunters are radioactive. Whenever a sheet is photographed any pairing will show up quite clearly. Sickle-cell anaemia is just one of the rapidly growing number of disorders that can be diagnosed from this kind of photographic evidence.

Existing and foreseeable medical techniques will do more than add new serious disorders to the list of prenatally detectable disorders. All sorts of *minor* defects will be detectable in the first and second trimesters of pregnancy.

> The once seemingly science fiction idea of routine, detailed fetal analysis in all pregnancies now is upon us . . . At present there are no laws, or even generally accepted guidelines, as to which disorders are appropriate for prenatal diagnosis and elective abortion. Technology is the only real limiting factor. There will be instances where technical possibility will be the major determinant for study rather than the consequences of the disorder. The development and application of these techniques will be by geneticists and others in related medical fields. The decision regarding how these techniques are to be used should not be left entirely to their discretion, however. It is hoped that prudence and wisdom combined with detailed, non-directive genetic counselling of couples will continue to be fundamental ingredients of prenatal diagnosis.[10]

Another once seemingly science fiction idea, that of very early embryo analysis, is discussed briefly in Chapter 7.

5 Abortion

*As conception is better understood and controlled – as the aura of
mystery around fertilisation and ovum transport is dispelled – when a
carefully prepared embryo transplant can replace an aborted fetus –
when abortion clinics become fertility control centres in the true sense,
with sperm banks as an integral part, society may finally resolve its
many conflicts over abortion and sterilisation.* (Jane E. Hodgson)[1]

Deliberate abortion may be almost as old as humanity. Anthropological studies on primitive tribes in isolated areas of the world suggest that prehistoric peoples induced abortion. Descriptions of abortion techniques appear in early Chinese, Egyptian and Greek texts. The 'Shen-Nung pên ts'ao ching', attributed to one of China's earliest legendary rulers, is said to have recommended mercury as an abortifacient. The Ebers Papyrus of Egypt (thought to date from about 1550 BC) gives two or more prescriptions for inducing abortion. And an ancient Hippocratic or pseudo-Hippocratic text prescribes violent exercises. The so-called Hippocratic Oath acknowledges the early existence of abortifacients by forbidding physicians to use them.

Plato and his pupil Aristotle supported abortion. Plato declared that men and women who had passed the best age for having children should be very careful that no child, if one happened to be conceived, should see the light. Aristotle called for a law that no deformed child should live. He did not recommend general infanticide: 'As to exposing or rearing the children born, let there be a law that no deformed child shall be reared; but on the ground of number of children, if the regular customs hinder any of those born being exposed, there must be a limit fixed to the procreation of offspring, and if any people have a child as a result of intercourse in contravention of these regulations, abortion must be practised on it before it has developed sensation and life; for the line between lawful and unlawful abortion will be marked by the fact of having sensation and being alive.'[2]

In ancient China abortion generally seems to have been regarded more as the parents' self-punishment than as a sin or a crime. When, in the nineteenth century A D, legal prohibitions did come in, this was due partly to the fact that Western powers had made reform of the Chinese legal system a precondition for abolishing the principle of extraterritoriality.[3] The Japanese may not have regarded abortion as a crime until the Meiji Restoration. Under the Emperor Meiji (1869–1912) a penal code was introduced with anti-abortion provisions on the French model. Sanctions in parts of Japan may have been more draconian on paper than they were in practice. The imported view of God as the Creator of Life competed with only limited success against Confucian philosophy.

In the Hebrew version of Exodus 21 a man who accidentally causes an abortion expects the death penalty only if the woman dies. The Greek Septuagint translation went further, exacting life for life if the embryo was 'formed'. Not every early Christian regarded deliberate abortion before the soul is in the body as murder.

Under the law of the later Roman empire the fetus did not enjoy the same status as a child or adult; treated as a part of the mother's body, it could be removed (provided the mother survived) without any risk of a charge of homicide. Removal without the husband's consent was a wrong to the husband. In practice, legal constraints on would-be aborters and their physicians were probably negligible throughout the greater part of the Roman empire's existence. Soranus' *Gynecology*, which may well have been written in Rome during the rule of Trajan and Hadrian (between A D 98 and 138), reflects a variety of moral attitudes. Some persons evidently believed that abortion was never acceptable; others held that abortion should not be resorted to lightly – it should never, for example, be performed in order to maintain youthful feminine beauty and it should not be used to conceal adultery. Some individuals reserved their approval for cases where the characteristics of the mother's uterus (or some other difficulty) meant that it would be dangerous to allow the pregnancy to go to term. Although the ancient practice of exposing unwanted newborn babies was made criminal under Roman law in

AD 374, it still survived in the sixth century (and very much later).

According to Muslim tradition the Prophet Mohammed began to receive the koranic revelations in or about the year AD 610. Sūra 75 of the Koran attempts to dispel scepticism about the resurrection (the power to give life to the dead) by pointing out that each of us was once a drop of semen which then became a clot of blood which was then formed and moulded by Allah, so that we became either male or female. In the Hadith (Al-Nawawi's Forty Traditions) the droplike state lasts forty days; then comes the clot and the flesh, and only 'thereafter' does an angel supply the spirit. The Grand Mufti of Jordan stated in 1964 that 'It is permissible to take medicine to procure abortion so long as the embryo is unformed in human shape.'[4] The unformed state was said to last 120 days.

Papal bulls and decretals have opposed abortion, but the amount of opprobrium varies. The thirteenth-century decretals of Pope Gregory IX contain a canon which treats abortion at any stage as homicide; the same decretals also contain Innocent III's opinion that abortion of a mere fetus should be distinguished from abortion of a vivified fetus. Graded penances for abortion are a feature of several collections of canon law. A collection of the tenth century exacted one year's penance if the fetus was under 40 days of age, three years' penance if it was over 40 days and ten years' penance (the penance for homicide) if animation had taken place. Pope Sixtus V, in the bull *Effraenatam* of 1588, invoked the legal penalties of homicide against all abortion, even where the fetus was unanimated or unformed. Aborters were to be excommunicated. Absolution from excommunication was to be reserved to the Holy See. Sixtus's successor, Gregory XIV, abandoned this policy in 1591, repealing all *Effraenatam*'s penalties except those applying to an animated fetus or ensouled fetus. This meant that any excommunicated aborter of an animated or ensouled fetus could seek absolution locally – there was no need to apply to the Holy See. In 1869 Pope Pius IX's Constitution *Apostolicae Sedis* effected a further change. Excommunication was made the penalty for all abortion; the formerly crucial distinction between animated and unanimated fetuses was abandoned.[5]

Under eighteenth-century English common law it may not have been a crime to abort (with the mother's consent) a fetus that had not quickened in the womb. The common law misdemeanour of abortion seems to have applied only to 'quickened' fetuses. An early nineteenth-century statute made it a felony punishable by death to administer a poison with intent to procure the miscarriage of a woman who was quick with child, and a felony punishable with either transportation for fourteen years or imprisonment to administer a poison with like intent to a woman who was not proved to be quick with child.[6] Later, the Offences Against the Person Act, 1828 extended these provisions to cover the use of any means (not only poison) to procure a miscarriage. The statutory distinction between quick and non-quick women may have led to difficulties in practice; it was abandoned by 1837 legislation which sought to re-enact this part of the criminal law in a more workable form. No such distinction appears on the face of any of the relevant statutes in force today.[7]

In the American colonies and the United States, the common law position in respect of abortion was similar to the position in England – abortion of a 'quick' fetus seems to have been distinguished from abortion with the mother's consent before quickening had occurred. In 1821 the General Assembly of Connecticut passed a law restricting the administration of a noxious or destructive substance to any woman who was quick with child. New York State legislation of 1828 prohibited surgical abortion even before quickening. A substantial number of states had passed restrictive legislation by 1850. Contemporary abortion techniques were notoriously dangerous and the primary concern of many a restrictive legislator was not ensoulment (of the embryo or fetus) but rather the life and health of the mother. By 1900 legislation had ensured that abortion was illegal in most American jurisdictions. Drafters of the restrictive laws allowed a defence where the continuance of a particular pregnancy would have threatened the life or (in some states) the health of the woman. Generally, however, the old 'quickening' rules had lost much of their earlier importance.

The Supreme Court of the United States ruled in 1973 that any state criminal abortion statute that excepted from criminality only a life-saving procedure on behalf of the mother, without regard to the stage of pregnancy and without recognition of the other interests involved, was violative of the Due Process Clause of the Fourteenth

Amendment. Whilst acknowledging that a state might have compelling reasons for regulating abortion, the Court ruled that it was unconstitutional for legislation to impose blanket prohibitions that had no regard to the stage of pregnancy:

> For the stage prior to approximately the end of the first trimester, the abortion decision and its effectuation must be left to the medical judgment of the pregnant woman's attending physician.
>
> For the stage subsequent to approximately the end of the first trimester, the State, in promoting its interest in the health of the mother, may, if it chooses, regulate the abortion procedure in ways that are reasonably related to maternal health.
>
> For the stage subsequent to viability, the State in promoting its interest in the potentiality of human life may, if it chooses, regulate, and even proscribe, abortion except where it is necessary, in appropriate medical judgment, for the preservation of the life or health of the mother.[8]

In the late 1970s there was reckoned to be one deliberate abortion in America for about every three live births. The current annual total of deliberate abortions in the United States may be of the order of 1.5–2 million. Perhaps 60 out of every 1,000 American women have had an abortion by the age of 18.

Under existing English law it is not murder to kill either a quickened or a non-quickened fetus in the womb. Current English statute law on abortion and child destruction can be found in the 1861 Offences Against the Person Act, the Infant Life (Preservation) Act, 1929 and the 1967 Abortion Act. Section 58 of the 1861 Act provides that 'Every woman, being with child, who, with intent to procure her own miscarriage, shall unlawfully administer to herself any poison or other noxious thing, or shall unlawfully use any instrument or other means whatsoever with the like intent, and whosoever, with intent to procure the miscarriage of any woman, whether she be or be not with child, shall unlawfully administer to her or cause to be taken by her any poison or other noxious thing, or shall unlawfully use any instrument or other means whatsoever with the like intent', shall be guilty of an offence.[9] Section 59 of the same Act makes it an offence to unlawfully supply or procure the means to commit an offence under section 58.[10] Since sections 58 and 59 apply only to 'unlawful' acts it was open to the courts to exempt, as lawful, abortions whose purpose was to preserve the life of the mother. Some lawyers thought that abortions performed in order to protect the mother's health were lawful.

Section 1(1) of the Infant Life (Preservation) Act, 1929 provides that 'any person who, with intent to destroy the life of a child capable of being born alive, by any wilful act causes a child to die before it has an existence independent of its mother, shall be guilty of [an offence], to wit, of child destruction'.[11,12] A crucial proviso says that a conviction under section 1 is impossible unless it is proved that the act which caused the child's death was not done in good faith for the purpose only of preserving the life of the mother. The 1929 Act states that evidence of twenty-eight weeks or more of pregnancy is prima facie proof that a child is capable of being born alive. Some babies are now being born alive before twenty-eight weeks, as technology and expanding medical knowledge transforms the care of the premature infant in medically advanced nations, and the Lane Committee, in a report published in 1974, wanted Parliament to impose a definite upper time limit for abortion of twenty-four weeks' gestation, so that thereafter every effort should be made to preserve the life of the child.

The 1967 Abortion Act creates important exceptions to the Offences Against the Person Act. A person is not guilty of an offence under section 58 (or section 59) of the 1861 Act when a pregnancy is terminated in an approved place by a registered medical practitioner, if two registered practitioners are of the opinion, formed in good faith, that the continuance of the pregnancy would involve risk to the woman's life, or of injury to the physical or mental health of the woman or any existing children of her family, greater than if her pregnancy were terminated, or that there is a substantial risk that if the child were born it would suffer from such physical or mental abnormalities as to be seriously handicapped. In determining the possibility of risk to health, account may be taken of the woman's actual or reasonably foreseeable environment. If there is an emergency and a registered medical practitioner is of the opinion, formed in good faith, that termination is immediately necessary to save the life or to prevent grave permanent injury to the mental or physical health of the woman, then the requirements as to the opinion of two practitioners and the place in which the operation is to be performed do not apply.[13]

Nothing in the Abortion Act affects the provisions of the Infant Life (Preservation) Act.

Of the 112,000 or so legal abortions that were performed in England and Wales in 1978, about 84 per cent were performed because of danger to the physical or mental health of the mother; 14 per cent were performed because of risk of injury to the physical or mental health of existing children of the pregnant woman's family; and 2 per cent were performed because of the existence of a substantial risk of the child having a serious handicap. Roughly 80 per cent of abortions take place in the first thirteen weeks, 18 per cent occur up to twenty weeks and 2 per cent occur between twenty and twenty-eight weeks. Failure to recognise pregnancy is a factor in more than 30 per cent of 'late' abortions. Apprehension and indecision on the part of women who are aware of their pregnancy may be a contributory factor in about half of the late abortions. Another factor, which may diminish, has been avoidable delay in the National Health Service. A recent survey of more than 600 doctors in England and Wales showed that 1 in 5 women medically referred for abortion before the twelfth week of pregnancy did not have the operation until between the twentieth and twenty-third weeks. It is not surprising that many women choose to pay for a private operation.

The maternal risks in abortion are significantly lower in the first trimester of pregnancy than in the second and third trimesters. If a woman takes immediate action on finding that her period has not started on the day that it was due, extraction of any product of conception from the womb by vacuum aspiration (suction) is a relatively safe and simple matter. Early in the actual or suspected pregnancy an extremely narrow tube (of about 4 mm in diameter) attached to a special 50 ml syringe may be all that is needed, and since such a tube can be introduced into the womb without dilation of the cervix it may be unnecessary to employ anaesthesia. A little further into pregnancy, up until the eighth week of gestation, vacuum aspiration is still the preferred technique in some British and American medical centres. Depending on the exact timing and the physiology of the patient, the diameter of the tubing will have to be slightly greater, perhaps 6 or 8 mm. Only minimal dilation of the cervix is necessary. After the eighth week, up until the twelfth week, vacuum aspiration of the womb's contents is still possible. The increased size of the embryo or fetus means that larger tubular instruments must be used. With tubes of 12 or 14 mm diameter, cervical dilation is necessarily greater.

Linda Francke underwent a successful first trimester abortion by vacuum aspiration in 1973 and she subsequently interviewed teenagers and fellow countrywomen who had experienced first or second trimester abortion in the United States. She concluded that 'There is . . . some discomfort both during and after the abortion procedure, the discomfort increasing with the length of the pregnancy. Abortion hurts. It is not unbearable pain, nor anything like the pain of childbirth (except in second-trimester abortions), but everyone suffers somewhat. After the abortion, many women experience cramps, which can be mild or so heavy that they necessitate pain medication. There is also the risk of hemorrhage or infection . . . I didn't bleed at all for two days just as they had predicted, and then I bled only moderately for another four days. Within a week my breasts had subsided and the tenderness vanished, and my body felt mine again instead of the eggshell it becomes when it's protecting someone else.'[14]

After about fourteen weeks' gestation, abortion by vacuum aspiration is no longer an automatic option. The fetus may be too big to pass through even a large-bore tube. This far into a pregnancy dilation of the cervix supplemented by the use of forceps and sharp curettage may be considered. The cervix is dilated sufficiently to allow the operator to scrape the womb. Dismemberment of the fetus is only required in the most extreme cases.

An alternative second trimester technique is intra-amniotic injection of an abortifacient such as hypertonic saline. Injected into the amniotic sac, saline (in common with urea and prostaglandins) initiates a shortened labour, persuading the uterus to empty itself. Hours may elapse between the introduction of the abortifacient and its effects on the uterus: the time interval varies from patient to patient and is not easy to predict.

Some operators prefer to use an extra-amniotic method for inducing labour. Instead of introducing an abortifacient directly into the amniotic sac the dosage is administered using an intravenous infusion or a pessary.

Women wanting to know what an abortion would be like sometimes consult friends or relatives who have had one. Useful information can be gained in this way, although allowance has to be made for the fact that some operations performed ten or even five years ago differ significantly from modern practice. Operational techniques are continually changing in response to new data and new possibilities. Clinical trials in the United Kingdom will test the effectiveness and safety of the French abortion pill R U486, said by its advocates to be usable up to about the eighth week of pregnancy. Termination of far advanced pregnancies is not yet (and may never be) an entirely straightforward and invariably simple matter but abortion can be very much less unpleasant for the woman, in physical terms, than it was just a decade ago, when medical knowledge of the side-effects of various procedures was more limited.

Writing about abortions undergone for reasons which had nothing to do with the diagnosed or suspected state of the fetus, Francke found that:

> Emotionally there is almost always a feeling of relief after an abortion, followed by a period of depression aptly named the 'post-abortion blues'. No one really knows what causes it; it may be as simple as the hormonal change the female body undergoes as it passes from being pregnant to not being pregnant, very much like the postpartum blues. The depression may be more severe in women who have been ambivalent about making the decision to abort, a situation that can be greatly alleviated by counselling both before and after the abortion. But for most women, any depression or regret fades somewhere between two weeks to six months after the procedure. The important thing to bear in mind is that having an abortion is not as simple as some of its advocates have led women to believe. It is a shock to the system, the womb in particular . . . Men . . . should know that after an abortion not only may a woman feel anger and hostility toward them, but that they

may suffer from the same emotions to such an extent that
they become temporarily impotent. And everyone should know
that, in spite of the rhetoric from the right, claiming that
women who abort are murderers, and from the left, claiming
that abortion is an instant panacea for an unwanted pregnancy,
the abortion experience is actually a period of great stress for
every person involved. There is indecision; there is pain. There
is regret, and there is relief. And all persons entering or leaving
the abortion experience without recognising the probability of
these emotions are simply fooling themselves.[15]

Elective abortion because of the actual or suspected state of the fetus

Linda Francke's research was concerned with abortions performed
for psycho-social indications. Although many of her observations
apply to abortion generally, it would seem that elective abortion on
eugenic grounds raises special issues.

In some X-linked conditions the decision to terminate is taken on
the basis of fetal sexing – science can establish the sex of a given fetus
and science can provide a risk estimate on the basis of the indicated
sex, but science cannot say that the fetus definitely has the disorder.
Where male fetuses with a 50 per cent chance of having a recessive
X-linked disorder are aborted, the parents, and especially the mother,
may be acutely vulnerable to the sort of critic who asserts that the
pregnancy would have given rise to a beautiful healthy little boy.
Such evenly balanced probabilities are not exclusive to X-linked
disorders for which no prenatal diagnostic test exists. Carriers of
certain autosomal dominant genes can face comparable though
much earlier decisions: Mendelian principles suggest there is a 50–
50 chance of an affected fetus in every pregnancy, irrespective of
the fetus's sex.

When a prenatal diagnostic test of high specificity and sensitivity is
available, the dilemma of 50–50 risk disappears. No longer will it be
as likely as not that a perfectly healthy fetus has been aborted. Few, if
any, clinical tests achieve 100 per cent accuracy in practice, but a
growing number of tests are at least 90 per cent reliable, and this
order of accuracy means that the elective aborter can be far more

confident, on statistical grounds, of the correctness of his or her identification – it's very much more likely that the aborted fetus did have a feared genotype than that it didn't.

In the early years of selective abortion on eugenic grounds quite a bewildering array of psychological studies were published. Some investigators made notably imprecise use of words such as 'guilt', 'depression' and 'regret'. Some investigators took little or no account of cultural differences over space and time. Recent critics of much of the early methodology have not been quick to draw up their own flawless and enlightening research programmes. It is still difficult to find clear, confident and uncontroversial pronouncements on the psychological implications of eugenic abortion.

More than one doctor in the 1960s reported that American women who had had abortions because of exposure to German measles (rubella) and the risk of consequential fetal abnormality reacted with depression, self-reproach and guilt. As a class these women apparently displayed a higher incidence of emotional side-effects than women who had had abortions for psycho-social indications. A team of researchers wishing to test the hypothesis that elective aborters of fetuses with an actual or suspected genetic abnormality have much in common, psychologically speaking, with women who abort in order to avoid the risks associated with German measles studied thirteen American families in the mid 1970s.[16] All of these families had experience of at least one abortion performed for a genetic indication after amniocentesis. One of the women and two of the men had not experienced post-abortion depression. Among the others, the intensity and duration of post-abortion depression varied enormously. There was, however, more depression than the team would have expected to encounter in a group of psycho-social aborters. And one internal control, who had experienced both an 'abortion of convenience' and a eugenic abortion, said that she had come to regard the latter operation as the worst thing that had ever happened to her. Interestingly this judgement was not formed until after she had failed in a subsequent attempt to produce a healthy child.

Among the group of thirteen families there were three sub-groups representing recessive, X-linked and sporadic modes of inheritance. Depression didn't appear to be greatly influenced by the mode of inheritance but, in an era when X-linked 'diagnostic' tests were based wholly on fetal sexing (50 per cent risk for a boy and 0 per cent risk for a girl), guilt feelings arising from the genetic situation seemed to be particularly associated with the women who were known to be carriers of an X-linked disease. In contrast, families coping with recessive or sporadic threats professed a sense of misfortune rather than feelings of guilt when talking about their role as transmitters of genetic disease. Autosomal dominant inheritance was not studied.

Theories about the causes of lengthy and intense depression in the wake of a eugenic abortion are plentiful. Following a prenatal diagnosis or fetal sexing which puts a particular pregnancy under a cloud, women and their partners have to face the fact (or the likelihood) that their attempt to produce a low-risk child has failed. Shamed by their inability to produce a 'normal' child on this occasion they also find themselves responsible for deciding what should happen to their creation. Even if there is no hesitation in deciding that abortion is in the best interests of the fetus and existing members of the family, there may be doubts about the acceptability of starting another pregnancy and risking similar disappointment. One investigator has suggested that depression is likely to be less severe in cases of non-recurring disorders, when couples can anticipate another pregnancy under low-risk rather than high-risk conditions.

Some women opting for a eugenic abortion look on the bright side. See Chapter 8, pages 136–8.

6 Contraception and Sterilisation

A contraceptive differs from an abortive, for the first does not let conception take place, while the latter destroys what has been conceived. (Soranus) [1]

Contraceptive practices down the ages have taken a variety of forms. Egyptian papyri dating from about 1900 1100 BC reveal details of semen-blocking preparations. Several recipes have survived. We can only guess at the relative popularity of pulverised crocodile dung and honey (to be used as a vaginal plug) and a vaginal pessary of lint soaked in the juice from acacia tips, with honey mixed in. Although these preparations preceded detailed biological knowledge of the respective roles of sperm and eggs by about 3,000 years, they must have offered quite a good chance of contraceptive success. So too did Casanova's legendary eighteenth-century technique of applying half a lemon to the cervix. Aided by sperm-killing creams and jellies, which are safer and more efficient spermicides than either lemon juice or the gum arabic that is contained in acacia tips, today's barriers such as the diaphragm (Dutch cap), cervical cap (check pessary), disposable vaginal sponge, vimule and vault cap all work in much the same way: they blockade the cervical opening of the womb and make it very difficult for active sperm to ascend the cervical canal and gain access to the tubes that connect the womb to the ovaries. The British Family Planning Association reckons that the modern diaphragm or cervical cap or vimule or vault cap is a very reliable method of birth control when used in conjunction with an appropriate modern spermicide. Of a hundred women using the method, between three and fifteen will become pregnant each year. Sponges containing a modern spermicide are less efficient, perhaps. The BFPA would expect between nine and twenty-five users out of every hundred to become pregnant each year.

The early European condoms, worn over the erect penis, were probably made from linen or the intestinal membranes of sheep. Fallopius describes linen and lamb's gut sheaths in the sixteenth century. Originally their primary use may have been as protection against sexually transmitted disease. After the industrial revolution had transformed much of the world, the development of the process of vulcanisation of rubber and then the replacement of rubber by liquid latex in the manufacturing process, allowed condoms to be used on a massive scale as a cheap contraceptive of quite high efficiency. Although hardly passion-proof, and best when used with a chemical spermicide, they have prevented countless unwanted pregnancies. Out of a hundred couples using the sheath with a spermicidal cream or jelly, there will be between two and fifteen pregnancies each year, according to the estimates of the BFPA.

The physician Soranus may have told his patients in the second century AD that 'Just as every season is not propitious for sowing extraneous seed upon the land for the purpose of bringing forth fruit, so in humans too not every time is suitable for conception of the seed discharged during intercourse.'[2] Yet it would be stretching a point to call Soranus an advocate of the 'rhythm' or so-called 'safe-period' method of contraception which is approved in Pope Paul VI's 1968 encyclical *Humanae Vitae* (see page 93 below). Human ova were first identified by von Baer in the 1820s. Not until about 1930 did Ogino of Japan and Knaus of Austria prove that ovulation (the release of an egg from one or other of the two ovaries) occurs between menstrual bleedings and that the stage the egg is released is relatively constant in relation to the next cyclical bleeding. What is now called the 'rhythm' or 'safe-period' method of contraception is based on the assumption that ovaries only release one egg per cycle; that each released egg is capable of being fertilised for only twenty-four hours or at most forty-eight hours; and that sperm deposited in the female genital tract lose their capacity to fertilise an egg after seventy-two hours. If all women menstruated regularly, with a cycle length of exactly twenty-eight days, and if every other assumption were invariably correct, the method would be more reliable than it is. Ovulation could be assumed to occur on, say, the fourteenth day and conception would be impossible if coitus was avoided just before, during, and just after the fourteenth day. In fact, menstrual cycles vary from month to month, sometimes by as much as thirteen days.

The BFPA estimates that of a hundred women who attempt to predict their 'safe days' on the basis of past calendar records, up to forty-seven will become pregnant each year. Women who combine diligent temperature recording with the detection of changes in cervical mucus, and other symptoms of ovulation, are very much less likely to have an unplanned pregnancy – of a hundred women using this method, between seven and fifteen will become pregnant each year. Many of the disadvantages of family planning using the 'safe period' could disappear if a new experimental urine test, designed to pinpoint when a woman ovulates, proves to be cheap, simple and effective.

Intra-uterine devices (IUDs) are placed above the cervix, inside the womb. Stone IUDs have long been used to prevent camel pregnancies. In humans, medically supervised development of chemically active and chemically inert IUDs is a more recent phenomenon. Though now of well-proven effectiveness, the exact way in which they work has yet to be elucidated. Unlike the barrier and spermicide methods, IUDs probably do very little to impede the progress of sperms. Instead, contraceptive success depends upon an interaction with the uterine wall (and possibly with the egg), which denies the fertilised egg an opportunity to become properly embedded in the lining of the uterus. The BFPA estimates that of every hundred women using an IUD, between one and four will become pregnant each year.

The contraceptive pill owes its existence to the discovery, made late in the nineteenth century, that certain hormones called oestrogen and progesterone are capable of inhibiting ovulation. It became possible in the 1950s to manufacture sex hormones synthetically and oral contraceptives were soon cheap enough to compete successfully with longer established methods of birth control. Several kinds of pill were developed. The progestogen-only pill prevents the normal cervical ovulatory mucus appearing, and by thickening the fluid at the neck of the womb makes it very difficult for sperm to swim through.[3] Changes in the womb lining, or endometrium, probably interfere with implantation of the egg. Additionally, there may be some hindering of egg release and transport down the tubes that connect the ovaries with the womb. The combined pill (progestogen and oestrogen) brings about the same kind of changes in cervical mucus as the progestogen-only pill and it affects the lining of the womb. Ovulation is usually suppressed and any egg that is released may have transport difficulties. The B F P A reckons that the combined pill is over 99 per cent effective, if taken properly. The progestogen-only pill is 98 per cent effective.

An intra-vaginal ring (I V R) which releases progestogen has been developed. The device is usually left in the vagina for a total of about three months, when a new ring is fitted. Early studies indicate that it may be about as effective a contraceptive as the I U D.

Intramuscular injection of progestogen can be a very efficient method of birth control, with cover lasting for up to three months after the injection. However the side-effects of this approach have caused some health professionals to regard the method as a contraceptive for exceptional situations.

Is the 'morning after' pill or injection a contraceptive measure or is it rather an attempt to procure a miscarriage? According to a 1983 statement by the Attorney-General made in written reply to a parliamentary question, the procedure does not constitute an offence contrary to sections 58 and 59 of the Offences Against the Person Act, 1861. Drugs which are administered after coition, with the intention of inhibiting implantation of a fertilised ovum, cannot properly be said to be administered to procure a miscarriage. The Attorney-General explained that the term 'miscarriage' in the Victorian Act referred to interference at a stage of prenatal development later than implantation. Some doctors believe that prevention of a fertilised egg's implantation into the womb wall, by whatever method this is achieved, is not abortion. Some clinicians are prepared to insert an I U D within forty-eight hours of any unprotected and regretted intercourse.

Since the 'morning-after' pill or jab exposes the woman to physical risks its commonest recommended use is for emergencies. Many family planning professionals in Britain approve its use in cases of rape where the victim was not protected against conception at the time of coition; and some doctors would like to see it used even more widely, following any unpremeditated intercourse that is regretted immediately.

Reference to male withdrawal before climax – *coitus interruptus* – can be found in Jewish, Christian and Muslim scriptures. Muslims allowed the method but the early Christian Church condemned it. The nineteenth century saw a revival of interest. Semen often seeps from the penis before climax; sperms move surprisingly quickly. Withdrawal is not regarded as a reliable method of birth control by the BFPA. Nor does the BFPA recommend the post-coital douche as a method of contraception. Ejaculated sperm can get well inside the womb within ninety seconds. Some doctors doubt the wisdom of washing out a woman's natural secretions; removal or dilution may encourage vaginal infections.

Pope Paul VI's encyclical *Humanae Vitae*, issued in 1968, says that: 'The direct interruption of the generative process already begun and, above all, all direct abortion, even for therapeutic reasons, are to be absolutely excluded as lawful means of regulating the number of children. Equally to be condemned, as the magisterium of the Church has affirmed on many occasions, is direct sterilisation, whether of the man or of the woman, whether permanent or temporary. Similarly excluded is any action which either before, at the moment of, or after sexual intercourse, is specifically intended to prevent procreation – whether as an end or as a means . . .'[4] Where there are sound reasons for spacing births, arising from the psychological or physical condition of the husband or wife, or from external circumstances, married people can try to 'take advantage of the natural cycles immanent in the reproductive system'. Not every Catholic has been convinced by the encyclical's suggestion that interference with the generative processes should be condemned as a crime against both God and nature. As Hans Küng has observed, 'Unconditional respect for nature results in a numinisation of it that conflicts with the modern view of human responsibility.'[5]

Sterilisation

Early forms of male sterilisation involved castration: a removal of the testicles. Punishment, the provision of *suprani falsetti* voices and the superintendence of female prostitutes were some of the reasons for the operation. Today, all routine operative methods are less radical. The aim is to block or remove only a short portion of the vas deferens as it lies in the upper part of the scrotum. The vas deferens is part of a transport system from the sperm-making testicles to the channel which runs through the centre of the penis. After a vasectomy the testes remain in place and continue to produce hormones and sperm. The sperm are unable to reach the channel which runs through the penis and they are reabsorbed internally. Since much ejaculate consists of secretions from the prostate and Cowper's gland, orgasm remains possible and it is accompanied by emission as usual. Even expert vasectomists cannot guarantee a 100 per cent success rate and although the operation is usually a minor one, performed under local anaesthetic and taking perhaps only ten minutes, there is some risk of complications. Successful vasectomies may or may not be reversible.

Female sterilisation usually involves an interruption of the continuity of the fallopian tubes, so that the egg cannot travel down them to meet a sperm. The fallopian tubes can be reached by several routes. When sterilisation is performed by laparoscopy (a technique in which the abdomen is inflated with gas to make the organs accessible and visible, and a delicate viewing instrument is inserted through the abdominal wall) the tubes can be fused or tied or have clips applied in twenty minutes or so. If the operation is performed under local anaesthetic and all goes well the woman may be able to leave after about an hour. Some surgeons prefer to use a general anaesthetic. This usually entails an overnight stay in hospital. The risk of complications is greater than in the case of vasectomy. The sex hormones which control ovulation and the menstrual cycle should not be affected. Only some operations are reversible.

In the United Kingdom voluntary sterilisation for social reasons has probably doubled for both sexes between 1974 and 1980. Concern about the side-effects of the pill, dislike of barrier methods of contraception and qualms about intra-uterine devices have contributed to the increased demand.

In the United States the proportion of voluntary surgical sterilisations carried out for non-medical reasons rose from 16 per cent in 1970 to 23 per cent in 1973. One study suggested that sterilisation was the most popular method of birth control among American couples where the wife was between 35 and 44; sterilisation was used by more than a third of the contraceptive users who were known to the researchers. General retrospective studies on the acceptability of voluntary vasectomy report that most vasectomised men (90–99 per cent) and their wives (88–90 per cent) view the operation favourably and would repeat the procedure. Less than 10 per cent reported that marital changes or infant mortality had made them regret the operation. This latter group did not, however, report significant harmful emotional or sexual effects. For historical reasons, the sequelae of laparoscopic tubal sterilisation in women are less well known.

In the context of coping with genetic risk, as distinct from maintaining family size at whatever level seems optimal in the absence of an exceptional genetic threat, the psychological sequelae of sterilisation may depend, at least in part, on the pattern of inheritance. Some carriers of harmful or potentially harmful genes feel personally obliged to undergo sterilisation themselves rather than have their non-carrier spouse subjected to the equivalent operation. At both the conscious and (some researchers believe) the subconscious levels there is sometimes a feeling that the person at fault should 'pay' for it. There may also be a rational argument – in the event of an accident or marital breakdown the non-carrier spouse might be the more likely to want to have children from any subsequent marriage.

Men can keep their options open if they bank some of their sperm before they are sterilised. Egg banks are not yet as common as sperm banks and it is not very easy to obtain large numbers of female gametes; but in principle women can also take steps to insure against a change of heart or changed circumstances. Artificial reproduction is discussed in the next chapter.

7 Artificial Reproduction

Be it enacted by the Queen's most Excellent Majesty, by and with the advice and consent of the Lords Spiritual and Temporal, and Commons, in this present Parliament assembled, and by the authority of the same, as follows:

1. – (1) Except with the authority of the Secretary of State under this Act, no person shall –

(a) procure the fertilisation of a human ovum in vitro (that is to say, elsewhere than in the body of a woman), or

(b) have in his possession a human embryo produced by in vitro fertilisation. (Enoch Powell's unsuccessful Bill)[1]

AIH and AID

In the eighteenth century a founder of scientific surgery, John Hunter, secured the insemination of a draper's wife by means of semen collected from the husband and injected into the vagina with a syringe. A child resulted. This early instance of medically supervised AIH (artificial insemination of a woman using sperm provided by her husband) was put in the shade a hundred or so years later by a group of American physicians. They successfully undertook AID (artificial insemination of a woman using sperm from a donor who is not her husband). Soon, both procedures came to be regarded as effective ways of treating certain types of infertility.

The *British Medical Journal* stimulated professional interest in artificial insemination by publishing in 1945 both a description of the technique and a favourable account of the physical results. The Most Reverend Geoffrey Fisher, Archbishop of Canterbury, responded to the unprecedented level of professional interest in the procedure by appointing a commission to consider artificial human insemination. When this commission reported in 1948 it criticised AID, and suggested that AID should be made a criminal offence. Archbishop Fisher told the House of Lords in 1949 that AID was wrong in principle and contrary to Christian standards.

A 1958 divorce suit in the Court of Session in Edinburgh brought AID to the attention of practising lawyers: under Scottish law, did AI with the seed of a donor who was not the married mother's husband constitute adultery? Extensive research, extending beyond Scotland's great legal writers, and encompassing the book of Deuteronomy, the writings of St Paul and the works of the Canonists caused the judge in the case, Lord Wheatley, to observe that 'The idea that adultery might be committed by a woman alone in the privacy of her bedroom, aided and abetted only by a syringe containing semen, was one with which the earlier jurists had no occasion to wrestle.'[2] Drawing attention to the earlier times when life was still regulated by the natural rather than the scientific order of things, an almost golden age when people knew what was meant by adultery, the learned judge concluded that the idea of *conjunctio corporum* was still an inherent concomitant of adultery. AID was not adultery.

Public controversy broke out in many parts of Britain, fuelled both by more or less accurate reports of the legal decision and by wild overestimates of the current numbers of AI births. During a debate in the House of Lords in February 1958 the Lord Chancellor assured the House that AI would be considered by a specially appointed committee. And later that same year a departmental committee did begin to 'enquire into the existing practice of human artificial insemination and its legal consequences', with the express purpose of considering whether, 'taking account of the interests of individuals involved and of society as a whole, any change in the law is necessary or desirable'.[3] Lord Feversham was the chairman. The Feversham Report of 1960 said that the practice of AID was disapproved of by a large section of society and that AID should be discouraged, even though some couples could benefit from it. There was a hope that the annual rate of perhaps 100 AID births would dwindle. At this time the annual total of medically supervised AID births in the United States may have been of the order of 5–7,000.

In 1982 the British Royal College of Obstetricians and Gynaecologists knew of over 1,000 pregnancies conceived and at least 780 live births following AID. Part of the increase in British AID births over the period 1960–82 should probably be attributed to a cautiously favourable British Medical Association Report (the Peel Report) which appeared in 1973, recommending that AID should be offered as part of the National Health Service.

Babies produced by artificial insemination do not seem to have more physical problems than babies who have been produced in the traditional way. Indeed some short-term studies suggest that the rate of birth defects may be a little lower. There is a need for long-term studies.

The success rate in terms of conceptions following AID is 50–80 per cent. Provided the donated semen is in good condition when it is deposited high in the vagina near the cervix or actually inside the cervical canal, and provided this operation is performed near the time of ovulation, with the inseminated woman remaining on her back for perhaps half an hour afterwards, to give the sperm easy access to the uterus, the chances of achieving a pregnancy in that cycle or, with appropriately timed later inseminations, within the next three or four cycles are good. Up to 75 per cent of women attending clinics for AID treatment will be pregnant after three months.

The semen is usually inserted with a syringe, having been obtained through masturbation. No surgery or anaesthesia is required. Semen mixed with a medium containing glycerol can be preserved almost indefinitely by freezing. Consequently it is no longer necessary to produce a fresh specimen before each insemination. Granted sufficient technological and administrative resources, stored semen may be used, after carefully timed and controlled thawing. This facility has led to the setting-up of sperm banks. The National Health Service in the United Kingdom uses a number of these. After considering evidence as to the health of children who have been created with frozen semen, the Warnock Committee of Inquiry into Human Fertilisation and Embryology concluded that frozen semen is not a major health hazard; the Warnock Report recommends that its use in artificial insemination should continue.

As the effectiveness of AID was being established, geneticists began to consider how the procedure might be used to assist couples with risky genes. Something like a quarter of the American physicians who practise AID are thought to have prevention of genetic disease as one of their aims. The figure for the United Kingdom is unlikely to be very far below this, and it may be even higher.

A male carrier of a deleterious autosomal dominant gene who is keen that his partner should bear children who will be free from the same gene may turn to AI, using semen from a healthy donor.

In disorders inherited as autosomal recessive traits, donated semen is sometimes the simplest bypass around a 1 in 4 risk in each pregnancy. Couples like Mr and Mrs Brown (see page 36) have used AID in order to avoid a recurrence of cystic fibrosis. If there is no reliable screening test to identify the carriers of a particular recessive gene it will be impossible to guarantee that even a healthy sperm donor is not a carrier. However, the odds against the donor having exactly the same potentially harmful gene as the husband has will be substantial. The odds in any given case will reflect the gene frequency in the relevant population. Whenever carrier detection is feasible, on the basis of a simple blood sample for example, all or much of the uncertainty about a healthy donor's genotype will disappear. Scientists now know the approximate position of the cystic fibrosis gene.

AID may also be considered in the context of X-linked, multifactorial and chromosomal disorders.

A national poll conducted in the United States in 1970 found that whereas 55 per cent of respondents approved of AIH, no more than 26 per cent approved of AID. Subjects of this poll were drawn at random from the general population and they did not constitute a high-risk group in genetic terms. A more recent American study of the attitudes of individuals who were at high risk showed a much higher level of support for AID; it approached 50 per cent. In the earlier national poll all kinds of reasons were given for rejecting AID. Some respondents believed that the procedure was morally wrong or constituted legalised adultery. Church disapproval was mentioned. So were: belief in Fate, preference for adoption, fear of marital friction, genetic considerations and lack of desire for a child that would be only partly the couple's own. Among the high-risk group, factors influencing respondents to utilise AID included avoidance of guilt and anxiety over the transmission of a lethal autosomal dominant gene, the wife's ability to contribute half an AI child's genes, her opportunity to experience pregnancy, joint control

over prenatal development and the anonymity of the AID procedure. High-risk respondents who had reservations sometimes gave
legal and religious complications as a reason but some viewed AID
as a threat to masculinity or doubted whether the husband would
feel like a real father; and others were not very keen on a genetic
input from an anonymous donor.

When AID is medically supervised the doctor often acts as an
intermediary, maintaining personal contact with the couple and the
sperm donor while preserving confidentiality to ensure that neither
they nor society know all the details.[4] The donor entrusts his specimen of semen to the doctor and the couple trust the doctor to find
a suitable donor. Just as adoption agencies have criteria for selecting
prospective parents so, too, do AI practitioners. These vary from
country to country, and from one practitioner to another, but some
or all of the following points are likely to be considered:

> strength of the couple's motivation;
> the husband, if not sterile or subfertile, or possessing
>   rhesus incompatibility, must have, or be known to
>   be at risk of transmitting, adverse genetic factors;
> the wife must be free from genetic disease and be able
>   to care for the child;
> there must be no deeply ingrained fears or prejudice
>   about AID;
> there must be mutual understanding between the
>   spouses and a likelihood of mutual understanding
>   between the couple and child;
> there should be a good family environment;
> both partners must be prepared to give formal written
>   consent to the AID treatment before it begins;
> the couple should be willing to respect the donor's
>   anonymity.

This list is not exhaustive.[5] Unmarried heterosexual couples living
together in a stable relationship may be considered. And some AI
practitioners see no reason why single females or homosexual couples
should be denied AID.

When selecting the donor a medically-oriented AI practitioner is likely to be favourably impressed by the characteristics listed below:

good health;
good, stable personality;
proven fertility (perhaps a minimum of two healthy
    children);
happily married;
wife free from any history of repeated abortion;
family tree free from a history of congenital or medical
    disease;
physical resemblance to the husband;
not known to the couple;
willing to undertake not to trace or attempt to trace
    the AID couple and their children.

In the early years of AI, before there were any reliable methods of storing semen, the AI practitioner often had to make do with semen collected from any married or unmarried medical student who happened to be on duty at the appropriate stage of a client's menstrual cycle. Unfrozen sperm deteriorate quickly and it was rarely easy to ensure that a non-local donor would be to hand at the optimum time. The new ability to store semen means that AI practitioners are no longer dependent upon locally produced semen. It is possible for them to use specimens from donors they have never met.

In 1982 the English Law Commission considered whether legal provision should be made so that the AID child would be entitled to discover the facts about his or her biological parentage. The Commission found that under the then existing law and practice the child's genetic identity would probably remain concealed if he or she had been entered in the birth register as the natural child of the mother and the mother's husband.[6] As things stood, the mother and her husband were usually effectively free to decide whether or not to disclose the fact that an individual was an AID child. Even if they did decide to tell an AID child what they knew, they would not usually be able to say who the sperm came from.

The Commission considered that the argument in favour of a procedure giving the child the right to know the facts about his or her conception is essentially that a person has a right to know the truth about his or her origins. Under adoption law, for instance, adopted children are entitled, on attaining their majority, to discover the facts recorded on their birth certificate about their natural parentage. Does logic dictate that an AID child should have the same right? If the only fact which an AID child is able to discover from the birth certificate is that he or she is genetically not the offspring of the mother's husband, but of a sperm donor wholly unknown not just to the child but to the mother and her husband as well, then this right of access to the original birth records might confer little real advantage. Changing existing law and practice by giving the AID child the right to know the identity of the sperm donor would have the effect of depriving donors and donees of sperm of a durable cloak of confidentiality. Impressed by these and other considerations (including the quite narrow scope of the report in which these matters were discussed) the Commission did not immediately seek to confer on the AID child the right to know its biological paternity. The authors of the report realised, however, that some right of disclosure might be thought appropriate if and when AID practice became regulated by statute.

In the summer of 1984 the Warnock Committee recommended that legislation be enacted to permit the husband of a woman artificially inseminated with donated semen to be registered as the father. The Committee also wanted Parliament to provide AID children who have reached the age of 18 with a right of access to basic information about the sperm donor's ethnic origin and genetic health. This second proposal would not give the AID child the right to know the identity of its genetic father. Nor does it amount to a right to obtain a description of the sperm donor's personality or life history. The Committee thought it was necessary to maintain the absolute anonymity of the donor. Anonymity served several purposes: it minimised the intrusion of an outsider into the family; and it protected the donor from the risk of being forced to accept parental responsibility at some later stage.

Information on the long-term social and psychological effects of AID is really quite hard to find. In the United Kingdom and in many other parts of the world, researchers intent upon identifying and questioning AID children face some formidable practical and ethical obstacles and it seems likely that a paucity of data will hamper rational discussion for many years to come.

The ethicist Joseph Fletcher has suggested that 'The fidelity of marriage is a *personal* bond between husband and wife, not primarily a legal contract, and . . . parenthood is a *moral* relationship with children, not a material or merely physical relationship. The claim that AID is immoral rests upon the view that marriage is an absolute generative, as well as sexual monopoly; and that parenthood is an essentially, if not solely, physiological partnership. Neither of these ideas is compatible with a morality that welcomes emancipation from natural necessity, or with the Christian ethic which raises morality to the level of love (a *personal* bond), above the determinism of nature and the rigidities of the law as distinguished from love.'[7] This passage was first published in about 1955. The purpose of AID is taken to be emancipation from infertile marriage. Married readers whose fertility is not in doubt but who are thinking of using AID in order to avoid a specific genetic risk must therefore decide whether they share the Fletcher view of marriage. Does support for AID in infertile marriage logically entail moral support for other categories of AID use?

When the Holy Office (Roman and Universal Inquisition) considered artificial insemination toward the end of the nineteenth century neither AID nor AIH was widely and openly practised. Asked 'Is artificial fecundation of women permissible?' the Office answered 'Non licere'.[8] Fifty years or so later Pope Pius XII informed the 1949 International Congress of Catholic Doctors that he condemned both AID and AIH if, in respect of AIH, the husband's sperm had been obtained through masturbation. Pius neither proscribed nor endorsed AIH following marital intercourse. More recently, in its evidence to the Warnock Committee of Inquiry, the Social Welfare Commission of the Catholic Bishops' Conference (England and Wales) has suggested that AIH in normal circumstances is quite acceptable. The Commission compared the procedure to assisting handicapped married partners to have intercourse with prosthetic devices: 'There is no bodily impairment to be "cured", but merely a functional disability to be overcome. There is no threat to the unity of the partners and the artificiality of the procedure need not be regarded in itself as morally, psychologically, or aesthetically significant.'[9] AID, however, was rejected by the Catholic Social Welfare Commission because the Commission believed that the probability of risk to the future happiness, security, and self-esteem of the child, and to the interfamilial relationship in which the child is involved outweighs the promise of the happiness of the spouses.

> A further concern would be the pressure for extension of AID for purposes other than the remedy of infertility. It might be claimed that it is a 'reasonable' step to demand AID in the case of a husband who could carry or transmit a genetic defect. From this, the next step would be to acknowledge the demand for AID in a situation where the male partner (or the female partner, in the case of egg donation) is genetically inferior in some not clearly defined respect. Bizarre though this might sound, it is clearly a likely development in a non-married situation that from the demand for a suitable/compatible donation can develop the demand for choice of the best possible. If selection is not left to the choice of the client, then the professional (or the clinic) is left with the choice. Here, with the demand for strict record-keeping, it is not difficult to imagine a clandestine, or informal, eugenics research programme, conducted with well-intentioned but misguided motives. After all, is the clinic staff to be ordered to choose at random amongst those donations not excluded?
>
> From the client's point of view, if there is clearly a wide choice of donors available, there will be pressure on clinic staff to find donations in accordance with the client's selective wish. What begins as a 'therapeutic' procedure only available within marriage, and to remedy infertility, or bypass a genetically transmissible defect, becomes viewed·as a eugenic benefit which others may demand inside and outside of marriage. It is foolhardy to dismiss such surmised changes in attitude, if the changes in attitude over the last twenty years towards abortion, or the letting die of newborn children, is recalled.[10]

Some men and women have no religious objections to AID but they would prefer to adopt because there seem to be fewer uncertainties at every stage of an adopted person's life than in the life of an AID child.[11] Legislators, social administrators, relatives and friends all seem to be more or less accustomed to the fact that adopted people exist. Even in jurisdictions where the law of inheritance, legitimacy and family obligation is being reviewed with adoptees in mind the scope for change seems small; citizens can expect a measured response to an already sizeable body of legal and sociological evidence. By contrast the myriad implications of AID are only beginning to be considered by specialists in all the relevant disciplines. In the United Kingdom and in most parts of the world, public education about the issues has hardly begun and readers can but guess at the future direction of official attitudes.

Under existing law neither AID nor AIH is unlawful. But a child born as a result of AID is illegitimate, liable to suffer every disadvantage that is associated with that status. According to the Warnock Report, the husband of the woman who bears an AID child has no parental rights and duties with regard to that child. These rights and duties lie with the donor who could, in principle, apply to a court for access or custody and who could, in principle, be made liable to pay maintenance. In practice, the usual anonymity of the donor means that it is rare for a donor to press for parental rights or be threatened with the burden of parental responsibilities.

The Warnock Report recommends that legislation be passed to make the AID child the legitimate child of its mother and her husband where they have both consented to AID. Furthermore, there should be a presumption in law that the husband has consented to AID, unless the contrary is proved. The semen donor should be deprived of all parental rights or duties in relation to the child. This proposal means that where it is shown that the husband has not consented to AID there will be no legally recognised father.[12]

Some potential parents reject AID not because of any qualms about the legal status of a future AI child, but because they believe that the distribution of deleterious genes should be left to Fate. The AI practitioner who regards AID as an excellent method of reducing the number of 'high-risk' births may be tempted to ask: 'How would you feel if you had to look into the eyes of a severely affected child who was brought into existence after you knew of the genetic risk and after the lower-risk alternative of AID had been explained to you?' But since individuals or couples with deep misgivings about AID are likely to experience exceptional psychological pressures

during the rearing of any AID child, there can be very few professional enthusiasts for AID who think it wise to bulldoze doubters
into using AID.

Men and women who do not view AID as an affront to genetic
fate and who have no religious or moral objection to the procedure
sometimes have doubts about the psychological implications for themselves, their partner, the sperm donor, the AID child and any friends
and relatives who may be told. Although the opportunity to avoid
feelings of guilt and anxiety over the transmission of a particular gene
or chromosome abnormality may be positively welcomed it is plain
that some potential users of AID worry about the possibility that the
genetically bypassed husband will find it difficult to love the child.
Advocates of AID point to evidence of strong 'paternal' love in some
stepfathers and suggest that this is even more likely to arise in AID
families because with AI the husband often appears to be in no doubt
that he is the child's father. Moreover, artificially inseminated women
have been known to say that their own early doubts about 'paternal'
acceptance have been groundless: 'He felt the child kicking inside me
and he watched her being born. I saw all the joy and pride of
fatherhood as he held his daughter and my worries about him disappeared.'[13] Writing of an AID boy, one 'social' father, who was
separated from the mother, said: 'My ex-wife and I are united in our
love for him . . . I am writing of the only human being I love more
than myself and it is very hard to be truly objective in describing the
qualities of my son.'[14] However, in this context, as in many others,
responses to infertility AID may be a very poor guide to responses in
AID undertaken for genetic reasons. The quite common practice in
infertility AID of mixing the husband's semen with the donated
semen is never followed when the object of AI is to eliminate any
possibility of the future child inheriting the husband's genes.

One couple attracted to AID said they preferred the procedure to
adoption because 'Any child we have will, at least, have a good
chance of inheriting some of the many good and lovable traits of my
wife.'[15] The genetics of personality traits are controversial and poorly
understood, even in scientific circles, but it would be surprising if this
sort of reasoning were altogether absent from the majority of decisions to use AID.[16]

For women who conceive through AID the pregnancy is no different in physical terms to ordinary pregnancies. The psychological and emotional rewards probably vary as much from woman to woman and from day to day as in conventionally achieved pregnancies. Many men and women who are willing to resort to AID as a bypass around genetic risk attach very great importance to bearing a child. Both husbands and wives talk of the woman's maternal instincts being satisfied in this way. In addition there seems to be a certain appeal in the prospect of doing everything that can be done for a child prenatally: couples can protect an AID fetus against the medical risks that are associated with mothers who continue to smoke during pregnancy; by taking special care over maternal diet a couple can reduce the risk of there being any congenital or developmental problem.

Is a child that is genetically the wife's but which is not the product of the husband's genes a potential source of stress for the wife? Will any such stress be compounded if the couple decide that the operation must be kept secret? Not every AID mother regards an AID child as more truly hers than her husband's; and it is not uncommon for the mothers of AID children to stress the fact that they and their husbands see AID children as products of the marriage – 'We see the child as ours.'[17]

Couples contemplating AID often wonder what the donor looks like, and their curiosity may extend to his character and intellect. In many cases this curiosity seems to dwindle once the AID process has been started. Some AI practitioners suggest that it is unlikely to be revived unless a particularly rewarding relationship with the first child prompts thoughts of a second AID child. When this happens a couple may exhibit a sort of brand loyalty, asking the AI practitioner to re-contact the donor or, in the case of a sperm bank with frozen assets, to see if there are any reserves. There may not be. Donors are frequently discouraged from producing more than half-a-dozen offspring as this reduces the chances of half-brothers and half-sisters meeting and procreating in ignorance of their true relationship. Limiting the productivity of donors in this way has the added benefit, some geneticists believe, of ensuring that his harmful recessive genes will not be spread disproportionately in the population. The

Warnock Committee wanted to see a limit of ten children who can be fathered by sperm donations from any one donor. The Committee envisaged a new centrally maintained list of NHS numbers of existing donors held separately from the NHS central register, so as to preserve the donor's anonymity. As a matter of principle the majority of the Committee did not wish to encourage the possibility of prospective parents seeking donors with specific characteristics. Accordingly, the majority wished prospective users of donated sperm to be kept in ignorance of the donor's character, his physical characteristics and achievements. Only a small minority of the Committee considered that there should be a gradual move toward the American practice of making detailed descriptions of the donor available to prospective parents, if requested, so that couples can make informed choices as to the donor they would prefer.[18]

However, the Warnock Committee did take the view that couples should be entitled to know the donor's ethnic group. Couples should also be provided with basic information about the donor's genetic health.

Occasionally a family donor is requested or insisted upon. It is not unknown in infertility AID for the wife's brother-in-law to be the donor. Such a person will sometimes be an extremely bad risk in AI that is undertaken to avoid the husband's genes. Nevertheless, family members are not incapable of being low-risk donors and, since affection and family ritual are likely to keep them in touch with the fruit of their seed, family donors may be exceptionally well placed to rebut the charge of irresponsibility which is sometimes levelled against the donor who creates life but then takes no interest in the child's development.

Is the role of the donor of such a kind that it is liable to appeal to the abnormal and unbalanced? The Feversham Committee asserted that this might be the case. Medical students who have actually given semen for AID tend to see things rather differently, likening themselves to the blood donor. A blood donor hopes to help people whose health or life is threatened by a deficiency which can be made good without physical harm to the donor. In infertility AID the donor hopes to provide a couple with the one substance they require. Is not the one kind of donor very much like the other? No. The semen donor, unlike the blood donor, is helping to bring into existence a new human being, and some commentators have suggested that important issues of personal and social responsibility turn on this distinction.

David Ison, when assistant curate of a London parish, remarked that in many human societies a man is responsible for all his offspring. This principle may be broken by illegitimacy but its existence is widely accepted. AID as commonly practised (and as the Warnock Report envisages it in the future) usually stands the principle on its head – the sperm donor is actually required to take no responsibility for the AI life he has helped to create.

> AID regards semen as a merely physical substance. A couple who want to overcome their infertility [or bypass a genetic risk] by AID may have some degrees of choice as to the characteristics of the donor involved in that function; it is up to them to invest this substance with their own emotional and personal significance. The semen is as impersonal as blood or sweat once it has been removed from its personal context. But can sperm be treated simply as a commodity in this way? Can it cease to have a connection with the donor? Giving semen is different from donating blood or bodily tissues: these are given to preserve a life which already exists, while the donor's sperm combines with an ovum to create a new life which owes half its physical origin to the donor, and inherits some of his physical characteristics. Whether the participants in AID like it or not, a physical connection between child and donor always exists, a connection which anonymity obscures but cannot break. The way in which the practice of AID uses impersonality to block emotional involvement in reproduction by the donor, while emphasising the personal nature of reproduction for the couple involved, simply goes to emphasise that reproduction is not merely physical, but encompasses the whole personality, and is an expression of that person. Personalities as well as bodies are involved. Once it is admitted that the process of reproduction involves the whole person, then the next question is whether a person can choose to make it impersonal, i.e. separate reproduction from its context of normal emotional involvement. Can the donor hand over his personal involvement in his

> physical offspring to the husband? Can the AID couple choose
> to see semen as impersonal and thereby avoid personal sexual
> input into their marriage from outside? The answers to these
> questions depend on the basic assumptions which are held about
> the nature of man.[19]

Nowadays adoptive parents are usually advised that early disclosure of an adopted child's status is in the child's best interests. But many AI practitioners advise couples not to inform the AID child of the circumstances of its conception. At one clinic in England, out of a total of 2,000 recipients of AID at that centre up to the spring of 1981, the staff apparently knew of only one couple who were intending to tell their children. In infertility AID long-term secrecy may well be the usual aim. There is a common fear that disclosure would produce adverse psychological reactions in the child and in other members of the family. Published reports of family development sometimes leave the reader to guess whether secrecy has been maintained throughout an AID child's adolescence and adulthood. When researchers report a mother as writing of her AID children that 'There is nothing that we would change in any way – they have given us such a happy family life . . . we think they have developed into well-balanced adults'[20] there is no way of knowing the extent to which the perceived success depended on secrecy or frankness. It is often quite hard to keep information from a child. Even young children are capable of sensing that adults want a particular topic avoided, and a child's awareness of heightened tension may precede any parental slip of the tongue.

When AID is undergone in families which are known to have a serious genetic disorder even short-term secrecy may be impossible to justify. The following letter by an AID child provides a thought-provoking perspective. The disease the writer's social father had is an incurable degenerative disease of the central nervous system which is inherited as an autosomal dominant trait (see page 40). All carriers of the Huntington's chorea gene are expected to develop the disease at some stage of their lives. The offspring of a patient have a 1 in 2 chance of inheriting the gene and the disease. The course of the disease is variable. At the time of the letter writer's conception pre-symptomatic heterozygote detection was impossible.

My mother conceived me by AID. My 'father' had, according to my mother, never shown any interest in sex or, indeed, physical contact of any kind . . . All this was alluded to me by the time I was four: in addition she said that as he was unable to 'put his seed into her, a doctor had put it there with a test tube', or words to that effect. I remember the discussion quite clearly for I wished to know exactly what a test tube looked like . . .

I knew that somehow my sister did not belong to 'us' – mum never described a pregnancy, birth, or breast-feeding with regard to my sister. By careful cross-questioning I extracted the information that my sister was adopted – she was told a year later. But never once did it enter my head that he whom we called my dad was not my real father. Grandmother [maternal] would note my resemblance to him whenever I lost my temper, neighbours would look me in the eyes, examine me minutely and remark how much I took after my father . . .

Mum decided that the truth must be revealed for I was developing, would soon be a woman. She was too embarrassed, I suppose, to tell me herself and she asked her lover of five years to explain the facts of my life to me . . . He sat me on his lap while mum stood anxiously behind. He asked if I knew about cows being 'injected' with test tubes of sperm – I said that of course I did, never being one to admit ignorance. There was obviously something more to all this but I still did not predict what was to come next. 'That's the way your mum fell for you.' There must have been more questions from me, answers from him but essentially I felt like escaping, carrying this great, shattering boulder of information away with me.

I had a picture of grunting farm animals, test tubes, sperm and me. God the Father had deserted me, I was the child of the devil: a pubescent melodrama that I acted out in hate and revenge.

In the months that followed I attempted to flesh out my biological father. My mother supplied the only information that she had. To make sure that his sperm was 'all right' every donor had to have three children. Somewhere I had more family. He had to be of good moral character, so I should be reassured that there was nothing of the criminal in his blood. He was probably a policeman, possibly a doctor . . .

Never did I worry about being a bastard. No, what upset my whole sense of being was that nobody knew my 'real' father: as though half of me did not, does not, exist.

But my mother clearly felt a sense of shame for I was sworn to secrecy. Therefore I told everybody the circumstances of my conception at the earliest opportunity. I sought out my most garrulous cousin and told her 'everything', in return for her juicy family secrets. She spread the news throughout my mother's family. To this day maternal aunts, who showed me affection in my early years, dislike and ignore me. Doubtless my personality has contributed greatly to this state of affairs but I do wonder . . . My grandmother always finds occasion to speak darkly about 'blood' in my company, blood and its mysterious capacity to 'carry' talents and traits. She always knew just how my mother had conceived me for she had paid the necessary fee for my conception.

My father died of Huntington's chorea four years ago. Mother's family commiserated warmly with my sister and all but ignored me. They knew logically I was no carrier of the dreadful disease but my 'father's' family have never been told. They avoided my eyes, pretended I did not exist. Thank God that an anonymous donor, with good blood, is my father and not a carrier of Huntington's chorea . . .

I married early and in the manner of the majority of adolescent girls who have no career, no alternative model, I longed for a baby. None came. After all the long-drawn-out tests we were told, or rather I was told, 'Your husband has no spermatozoa – he is sterile.' I was eighteen and the question of AID came to my mind in a new light. In discussions with my mother it came out that she had worried about . . . my origins, my father, from whom I had inherited all undesirable traits . . .

Subsequently I . . . came to understand my mother's decision to have a baby by AID. She was no thoughtless ogre but a woman who craved her own child. But me, how can I make such a decision? My child would lack two generations of fathers. I could not hide the circumstances of my own conception for it might well find out about my 'father's' inherited illness and be full of fear. Friends and relatives all know that my husband is sterile (he has had a biopsy to confirm that his condition is irreversible). Children sense the unsaid – it would have to be told

and told young. My husband and I have been unable to resolve
the dilemma. Instead I went to Teacher Training College and
now teach nursery children thus detracting from my desire for
children, not that this helps my husband in any way.[21]

External fertilisation

In 1977 the English obstetrician Mr Patrick Steptoe collected a mature
egg from a woman with blocked fallopian tubes. His colleague, Dr
Robert Edwards, fertilised the egg in a glass dish, using sperm from
the woman's husband. When this fertilised egg developed to about
the eight-cell stage Patrick Steptoe collected it in its culture fluid and
placed it inside the woman's uterus. Implantation and all the usual
stages of pregnancy followed, and Louise, the world's first IVF child
(the first child produced by *in vitro* fertilisation), was born in England
in the summer of 1978.

Six summers later, the worldwide total of IVF births was
numbered in hundreds. Scientists found that it is not all that difficult
for a gifted operator to recover one or more eggs from a woman's
ripe ovarian follicles. The original operation had been performed
with a laparoscope, an optical surgical instrument which is usually
used in conjunction with a general anaesthetic. Ultrasound can now
be used to identify the position of the most promising follicles.
Under local anaesthetic a needle is passed through the abdominal
wall, guided to its target under ultrasound vision and used to suck
out the egg or eggs. Once one or more suitable eggs have been
obtained it is a relatively simple matter to achieve fertilisation of the
human egg *in vitro* (inside a glass dish). The main practical difficulties
arise at the post-fertilisation stage.

In conventional reproduction, as we noted in Chapter 3, some
fertilised eggs are lost at a very early stage of their development. The
natural rate of loss means that the chances of achieving a successful
pregnancy with just one IVF egg or embryo are rather modest. A
much more efficient method of producing offspring is to transfer
several fertilised eggs at a time. It is possible to stimulate the ovaries
of a woman with drugs to ensure that she produces several eggs in
one cycle. If superovulated eggs are mixed with semen, a high

proportion will be fertilised. The precise number of fertilisations cannot be predicted. Nor do doctors have any means of foretelling the number of fertilised eggs which show early signs of abnormal development that mark them out as being unsuitable for transfer to the womb. Spare embryos – fertilised eggs which nobody intends to use for procreation – are not uncommon.

The world's first IVF child derived half her complement of chromosomes from the woman who gave birth to her and half from that woman's husband. Neither side of the family was thought to have an exceptional complement of deleterious genes. There was no genetic reason to use a donated egg or donated semen. In some families, however, the known or suspected presence of certain deleterious genes makes it exceptionally risky to use the woman's own egg in procreation. A donated egg, fertilised *in vitro* by the husband's or partner's sperm, might be safer.

In genetic terms, egg donation is simply the exact complement of AI by donated semen. Unlike AID, however, IVF using a donated egg is quite complex. If no use is made of frozen, stored and thawed eggs, then the donor (who may be a woman undergoing infertility treatment or who may be a fertile woman who has requested sterilisation) will have to provide ripe eggs at the right point of the recipient's cycle. Monitoring and co-ordinating all the relevant events is extremely time-consuming, and some practitioners look forward to the day when banks of mature eggs are as common as sperm banks. Treatment of infertility and the bypassing of deleterious genes in certain females would be easier still if immature eggs could be stored indefinitely and then be matured *in vitro* with perfect safety as and when they were required.

Where the original possessor of a donated egg is not the person who becomes pregnant and gives birth to the child, who then is the mother? English common law provides no obvious answer. The Warnock Committee hoped that Parliament would legislate so as to make the woman giving birth the legal mother; the egg donor would have no rights or obligations in respect of the child.

Members of the Committee acknowledged that egg donation is open to the same kinds of objection as AID. In egg donation a third party is introduced into the marriage (or other relationship). The impact on the child is uncertain. Possible harmful effects on society can be imagined. In addition, the procedure involves very considerable intervention in the normal process of fertilisation. Moreover, egg donation, unlike sperm donation, requires invasive procedures which expose the donor to a degree of physical risk. To be set in the balance against all these objections were the new opportunities some couples would have. A donated egg may provide a couple with their only chance of having a child which the woman can carry to term, and which derives half its complement of chromosomes from the husband. The Committee hoped that egg donation would come to be accepted as a recognised technique in the treatment of infertility and in the avoidance or diminution of particular kinds of genetic risk. The principles of good practice would be similar to the principles in sperm donation – limitation of the number of children born from the eggs of any one donor to ten, openness with the child about the child's genetic origins (falling well short of disclosure of the egg donor's identity) and the availability of counselling for all parties. In the first years of I V F it was almost impossible to devise procedures that would ensure that every egg donor remained anonymous. Now that scientists are developing methods of freezing, storing and thawing human eggs, anonymity will be relatively easy to arrange.

Embryo donation, the gift of a fertilised (rather than an unfertilised) egg, can take more than one genetic form. A woman may receive a donated egg that has been fertilised *in vitro* with semen from her husband or partner, but it is also possible to have the donated egg fertilised *in vitro* with semen from a man who is not her husband or partner. Where the donated eggs are obtained by minor surgery the egg donor undergoes some physical risks. Many of these risks can be avoided if the egg donor is willing to allow herself to be artificially inseminated at the predicted time of ovulation and to keep the fertilised egg inside her reproductive tract for some three to four days, before it is washed out by doctors, retrieved, and then transferred to the uterus of the woman who hopes to bear and give birth to the eventual child. No general anaesthesia is required for the artificially inseminated donor. Her principal risk is an unwanted pregnancy – the embryo may resist the doctors' attempts to wash it out. There is also a chance that infection will be introduced to the uterus.

The Warnock Committee believed that embryos should not be donated by the technique of A I and washing out. Lavage techniques in use at the time the Report was prepared (the summer of 1984) exposed the egg donor to unacceptable risks. However, if these risks were overcome, the Committee would withdraw its objection to the technique and would recommend that any embryo donated in this way should be treated in law as the embryo of the woman who carried the pregnancy. The Committee said that the form of embryo donation involving donated semen and egg brought together *in vitro* was already acceptable as a treatment for infertility and should be available to those at risk of transmitting hereditary diseases. Once transferred, any embryo created in this way should be treated in law as the embryo of the woman who carries the pregnancy.

Surrogacy

The practice whereby one woman carries an embryo and fetus for another woman, with the intention that the eventual child should be handed over after birth, can take several forms. A carrying surrogate mother need not be the genetic mother. The genetic father may be married to the woman who commissions the pregnancy or to the carrying mother or he may be an anonymous donor.

The majority of the members of the Warnock Committee recommended that legislation should be introduced to render criminal the creation or the operation in the United Kingdom of agencies whose purposes include the recruitment of women for surrogate pregnancy or making arrangements for individuals or couples who wish to use the services of a surrogate mother. The majority also hoped that such legislation would make it a criminal offence for professionals and others to knowingly assist in the establishment of a surrogate pregnancy. Privately arranged surrogacy agreements were not a target of the envisaged criminal legislation. Committee members had no wish to render private persons entering into surrogacy arrangements liable to prosecution; that would lead to children being born to mothers subject to the taint of criminality. The Report did, however, express the hope that Parliament would provide by statute that all surrogacy agreements are illegal contracts, unenforceable in the courts.

In March 1985 the Secretary of State for Health and Social Services told the House of Commons that the government believed that commercial surrogacy was undesirable in principle. He would bring forward a bill prohibiting commercial agencies from recruiting women as surrogate mothers and from making surrogacy arrangements. The bill would also prohibit advertising of these agencies' services.

The Secretary of State acknowledged that the question of surrogacy raised wide issues, both of general principle and of law. He told the House of Commons that the government thought it right to deal with these questions in the comprehensive legislation which was needed to deal with the entire range of issues raised in the Warnock Report.

Comprehensive Warnock legislation is still awaited. There is, however, a Surrogacy Arrangements Act. This prohibits both the recruitment of women as surrogate mothers and the negotiation of surrogacy arrangements by agencies acting on a commercial basis. It also prohibits advertising of or for surrogacy services. The Act extends to the whole of the United Kingdom.

Detection of sex, gene defects and chromosome abnormalities in early, unimplanted embryos

Techniques for fertilising human eggs *in vitro* may lead to the once seemingly science fiction idea of screening very early embryos for sex, gene defects and chromosome abnormalities. Having produced an embryo in a glass dish, scientists might be able to perform a vast number of diagnostic tests, thereby enabling the parents to make an informed choice about the wisdom of implanting it. Rejected embryos could be destroyed without the trauma of abortion.

Some scientists dream about the day when doctors will be able to accomplish therapeutic genetic change in an externally fertilised egg.

Other scientists would like to be able to act before fertilisation, screening gametes so that only the safest create new life.

8 Decisions

When you toss a coin nobody, nobody, can tell you, even the cleverest of people, which it is going to be – heads or tails – because it might be either. (*Anonymous*) [1]

Family trees and genetic registers

Few of us know our pedigrees in great detail. If a medical disorder occurs in our offspring or in some other less closely related member of the family it may be necessary for our doctor or specialist clinical geneticist to ascertain the age, sex and medical history of the patient's parents, siblings and other near relatives. Information about still-births and miscarriages may be required. The age at which relatives died and the causes of death can also be important. Direct questions may have to be asked about consanguineous marriages and non-paternity.[2]

Some couples will learn that there is no evidence that their planned but as yet unconceived offspring would be exceptionally likely to inherit a particular disorder. Other couples will be told that the risk of recurrence is not negligible; being estimated at, say, 10 per cent.

In his *Treatise on the Supposed Hereditary Properties of Diseases containing remarks on the unfounded terrors and ill-judged cautions consequent on such erroneous opinions*, which appeared in London in 1814, Joseph Adams, a pre-Mendelian apothecary-physician with a very enquiring mind, called for the setting-up of family registers. 'To lessen anxiety, as well as from a regard to moral principle, family peculiarities, instead of being carefully concealed, should be accurately traced and faithfully recorded, with a delicacy suited to the subject, and with a discrimination adapted to the only purpose for which such registers can be useful.'[3] Adams did not overlook our propensity 'to conceal the skeleton in the closet'.[4] He acknowledged that many people feared enquiries into the genetic truth. Men and women who are aware of a hereditary or supposedly hereditary disease in their family often conceal or deny the fact. Men and women with no knowledge of a genetic skeleton in the family sometimes resist any attempt to inform them of hereditary dangers.

Adams was not so foolish as to suppose that genetic registers were a panacea. He wondered, however, whether it might not be found that in uneasiness stemming from hereditary disease 'a more accurate knowledge is less painful than constant suspense'.[5] What, however, of persons who have never given hereditary disease a thought (and who aren't, therefore, suffering well-informed or groundless unease)? Should they and their families be investigated? Adams believed that genetic enquiry cannot be 'unimportant to any family, how free soever they may fancy themselves from any hereditary peculiarity; for it will require no argument to prove that, like the varieties in other animals, all these peculiarities must have originated in the offspring of couples who were free from them'.[6]

A century later, Charles Davenport of America's Eugenics Record Office lamented the fact that his contemporaries tended 'so readily to associate a surname with a trait, bad or good, and thus to handicap it unduly; or, on the other hand, yield to it a confidence that is not warranted'.[7] He himself believed that 'In marriage selection nothing can take the place of a careful consideration of the probable germ-plasmic content of the two persons involved.'[8] Some of Davenport's views on germ-plasmic content have been discredited. Today few, if any, scientists share Davenport's belief that traits such as criminality and tone deafness are monogenic. Yet Davenport made accurate as well as inaccurate use of Mendel's newly rediscovered insights into inheritance. Having studied the pedigrees of numerous American families Davenport interpreted a certain form of late onset disease as a unifactorial dominant trait. He said that half the children of every affected parent would become affected (on the average). There was no cure. The disease killed slowly. Protracted physical dependence and deteriorating mental functions were common features.

Davenport thought it would be a work of far-seeing philanthropy to sterilise every American citizen who showed symptoms of this hereditary disease. America's gene pool could be protected from outside contamination if the State would only insist on knowing what the parents or grandparents of each would-be immigrant had died of, or what diseases they were liable to, if still living. 'We think only of personal liberty and forget the rights and liberties of the unborn, of whom the State is the sole protector . . . the writer has often insisted on the duty of the nation to know something of the blood lines of its imported human stock, as it does of its imported cattle. Had it been known that one parent of the three brothers who came in the seventeenth century from England was choreic, and had they been excluded on that account, we should have lost two leading educators, a surgeon or two, two state senators, two or three State Assembly men and several ministers and 900 cases of one of the most dreadful diseases that man is liable to.' [9]

Davenport's profit and loss account has several curious features. The lives of literally hundreds of descendants who did not in fact inherit the gene are passed over in silence. And Davenport nowhere explains why the State should be regarded as the *sole* protector of the unborn. His own belief that this was the case, or that it ought to become the case, can have offered him little comfort – states (notably those which disregarded his advice) were quite capable of being 'stupid'. [10] Nevertheless, it was Davenport's fervent hope that the State would prevent individuals from creating new life if they were heterozygous for the autosomal dominant gene he feared.

Eugenicists are sometimes a little vague on human detail. Overworked and unable to give close attention to his publisher's proofs, Francis Galton for instance failed to correct an extraordinary commingling of his notes on the families of Jane Austen, the novelist, and of Austin the jurist. The mix-up was published. When Galton subsequently came face to face with a representative of each of these families in the Athenaeum Club they 'gave it me hot, though most decorously'. [11] Galton was fortunate to escape so lightly. The title of his errant work – *Hereditary Genius* – may have helped. We all make mistakes. [12] The errors of authoritarian eugenicists are especially regrettable.

Save in respect of 'mental deficiency' [13] – an elastic term which could be stretched to accommodate the whole of the 1926 Tory government (according to a Labour MP of the day) – compulsory sterilisation seems never to have been the major plank in the programmes of the London-based Eugenics Education Society.[14] Writing in 1912 the chairman of the Belfast branch of the Society expressed the hope that 'sterilisation of the "unfit", which is at present being practised, with dubious results, in several states of the American Union, will not be pressed. Even if it could be justified, which is doubtful, public opinion in this country is not ripe for so drastic a proceeding. The public conscience would be shocked by it, and a promising movement would probably receive a rude check. Many feel instinctively that we might purchase a biological benefit too dearly at the cost of a spiritual wound.'[15]

When a member of the Eugenics Society sought leave to introduce a eugenics bill in 1931 he told the House of Commons that his measure was intended to enable mental defectives to be sterilised upon their own application, or that of their spouses or parents or guardians; and for connected purposes. (At this time voluntary eugenic sterilisation was widely thought to be illegal.) The motion failed.

Quite similar proposals in the Brock Report of 1934 were stillborn. Subject to specified safeguards, the 1934 Report recommended that sterilisation should be legalised in the case of: mental defectives; actual or suspected cases of a grave transmissible physical disability; and persons believed likely to transmit mental disorder or defect.

The Brock Report noted that 'Germany recently made a law of a comprehensive character intended to prevent the transmission of hereditary disorders both physical and mental. It permits the voluntary sterilisation and also provides in certain circumstances for the compulsory sterilisation of persons suffering from congenital mental deficiency, schizophrenia, manic-depressive insanity, hereditary epilepsy, blindness or deafness and other heritable conditions.'[16] German applications for sterilisation were to be decided by tribunals with a legal president who could be outvoted by the two medical members that sat with him. The Brock Committee had very little information on the working of the new German law. Nazi eugenics with anti-Jewish intentions and authoritarian recourse to compulsory sterilisation in respect of a wide range of both congenital and sup-posedly heritable conditions helped to make 'eugenics' a very dirty word.

. In 1955 the American ethicist Joseph Fletcher asserted that 'The right of society to be clean and safe, and the right of every child to be sound of mind and body, are the things at stake in the argument for compulsory sterilisation. Developing research has tended to show that many diseases . . . are not only infective or environmental in origin but may be due to an hereditary vulnerability as well . . . It is impossible to see how the principle of social justice – at least on any very profound view of it – can be satisfied if the community may not defend itself, and is forced to permit the continued procreation of feeble-minded or hereditarily diseased children. Sterilisation in such cases is not solely a matter of commutative justice (or personal con-trol), but also of distributive justice (or State control).'[17]

Today, in both America and the United Kingdom, it is widely appreciated that no amount of sterilisation can ensure that every member of the next generation will have a sound mind and body. Genetic registers do exist, but they can hardly be characterised as sterilisers' hit-lists.

. When British registers were investigated late in the 1970s, it was found that almost forty British centres maintained registers that were primarily concerned with genetic disease. These registers fell into five categories. Twelve per cent of currently maintained registers had a clinical purpose – they were intended to make it easier to follow up and recall individuals for therapeutic reasons, or when a breakthrough with a relevant preclinical or prenatal test occurred. Seventeen per cent of registers were reference registers, maintained so that the diagnosis in a new case might be confirmed by reference to previous cases. Twenty-six per cent were monitoring registers, facilitating assessment of the results of genetic counselling and prenatal diagnosis. Thirty-two per cent were used for research purposes, so that the natural history of particular disorders could be traced and their dis-tribution analysed. Only thirteen per cent were dedicated to pre-ventive medicine, and none of the registers in this category was maintained in order to perform compulsory sterilisation. Prevention of new cases was attempted through improved ascertainment and follow-up of individuals at risk of transmitting a serious genetic disorder to their offspring, so that they might be offered professional genetic counselling and prenatal diagnosis where possible.

When a member of a large family is found to have a disease which is inherited as an autosomal dominant trait (see page 40) there will usually be many other family members at quantifiable risk of having affected children. The same is true of X-linked recessive disorders (see page 50). Genetic registers are used in every type of inheritance, but perhaps their potential to prevent future cases is greatest in the case of autosomal dominant and X-linked recessive disorders. Once the diagnosing physician or clinical geneticist knows the identity of his patient's relatives, the diagnosis can be used in conjunction with knowledge about the mode of inheritance to work out the level of risk for every family member. In other words, a single index case facilitates the ascertainment of many at-risk relatives. This sometimes occurs in multifactorial inheritance (see page 53) but, for reasons that were touched on in Chapter 2, the risk levels in multifactorial disease tend to be lower, where relatives of the index case are concerned, than in X-linked or autosomal dominant inheritance. Similarly, though for different reasons, it is rare for a single autosomal recessive case to place many family members at 'high' risk (that is, greater than 1 in 10 – see pages 130–32) of having an affected child.

Studies carried out in the early 1970s in Britain suggested that many of the individuals who were deemed to have a risk greater than 1 in 10 of producing a child with a serious genetic disorder were totally unaware of their risk. Only about 15 per cent had been referred for genetic counselling. This picture could obviously be transformed if computerised genetic registers were to become popular among both health professionals and members of the general public. There would then be less wastage of information. At present, if the at-risk relatives of an index case are below reproductive age when the key diagnosis is made, it is quite likely that the doctor who makes the diagnosis will lack the necessary resources to store the information, and to keep track of several changes of address, until the relatives have become old enough to be informed of the potential risks for any child they may be thinking of having.

When sensitive information is to be stored in a computerised register, the providers and potential providers of information are likely to consider two matters. Who will have access to the information? How will other family members feel when, as a result of their names being put on the register, they are informed of a genetic risk?

The issue of confidentiality is a difficult one. Many operators of computerised registers use a consent form. The reader might, for example, be asked to sign a piece of paper which reads:

> I   of   give the Department of Clinical Genetics at
> the   Hospital,   , my consent to record details of myself
> and/or my children in the Department's Genetic Register
> System. I understand that the information contains genetic and
> medical details; that the system is computerised; and that the
> register is strictly confidential, access being limited to the
> medical staff concerned with my condition.
>
> I also give the said Department my consent to contact my
> relatives about the genetic disorder in my family.[18]

The assurance that the register is 'strictly confidential' is important – confidentiality is what many patients expect in a doctor–patient relationship. The phrase about access being limited to the medical staff concerned with the signatory's condition is less clear-cut. Does it include medically qualified researchers the signatory has never seen? Presumably. Might it include medically qualified individuals who hold threatening views on preventive sterilisation? Doctors working in a clinical genetics unit tend to have an implicit belief in the individual's freedom of choice. Using information provided by the signatory, might a member of a hospital department make contact with a relative and, having no cure to offer, destroy that relative's peace of mind? That could happen. On the other hand, an unseen member of a medical genetics team might be able to banish a relative's unwarranted fears by tracing him or her and then demonstrating, through some newly discovered test, that he or she does not in fact have a dreaded family gene.

Attitudes to the acceptability of genetic registers vary. Some members of the public are relieved to learn that certain departments of clinical genetics do not contact any relative of an index case if the relative's family doctor opposes such a step. Other members of the public are indignant that risk information can be withheld on the say-so of a relative's practitioner.[19] Where the at-risk relative is old enough to understand, but his or her own doctor refuses to give consent to professional counselling, it sometimes happens that the index case and near relatives take matters into their own hands, by doing the counselling themselves. This may work brilliantly, with accurate information being accompanied by affectionate support from family members. Occasionally, however, do-it-yourself counselling becomes a tragic game of Chinese Whispers – avoidable wretchedness flows from misunderstood, misremembered and misquoted statements.

Professional counselling

Many specialist genetic counsellors describe their work as a communication process that deals with the human problems associated with the occurrence, or risk of recurrence, of a genetic disorder in a family. Using a wide knowledge of genetics, medicine and psychology, they attempt to help their clients to understand the medical facts, including the diagnosis, probable course and responsiveness to treatment. They also try to get their clients to appreciate the way heredity contributes to certain disorders and the risks of recurrence in specified relatives. They discuss the options for dealing with any recurrence risk and they encourage clients to choose the courses of action which seem appropriate to the clients in view of their risk, and their moral standards and practical goals. Clients are encouraged to act upon their choice and adjust as best they can to the disorder in an affected family member or to the risk of recurrence.

The process of assimilation may be extremely rapid, with a couple needing only a few minutes to extract what they need from the counsellor; alternatively more than one lengthy session may be required. Much depends upon the facts, dilemmas and personalities in each case.

Understanding requires attention, comprehension and retention. The mother who is still shocked by the realisation that her newly born child has a serious genetic defect or the man who is still shaken by the news that he has transmitted a serious flaw to one or more of his children may be quite unable to assimilate even simple information. Women who produce children with serious genetic disease often seem to experience stronger feelings of guilt than their partners; this may reflect a mother's closer feelings in respect of her children and the greater amount of time she spends with them, both at home and at clinics for sick or handicapped children. Some genetic counsellors suspect that counselling is more influential in alleviating feelings of self-blame in women than in men. Experienced counsellors try to recognise the often very different needs of individual family members.

After the birth of a child with a genetic disorder, the relationship between spouses is likely to change. There may be unprecedented closeness or extra strain. A single diagnosis can also transform relationships between siblings and between other persons who are related, whether by blood or by marriage. Not uncommonly, the grandparents of a seriously affected child are just as angry with their God or with the godless world as the child's parents are. Unaffected members of an affected sibship may believe that they have escaped Fate at the expense of their affected brothers or sisters. Dissimilar and changing family perspectives on the seriousness of the threat of repetition can result in muted or extravagant expressions of delight when a couple who are known to be at risk of producing children with a specific disorder announce a new pregnancy.

Some counsellees bring to their first appointment an excellent knowledge of genes, chromosomes, patterns of inheritance, probability, and the prognosis for existing and contemplated family members. 'I felt I knew as much as the genetic counsellor and that his role – valuable as such – was simply to reiterate what we already knew. There was not much doubt that the syndrome is caused by a recessive gene in both parents and that the chances of us having a similarly affected baby were 1 in 4. He was concerned and sympathetic and did discuss options open to us – adoption and AID. Having heard what our inclinations were he made referrals to appropriate people. There was no pressure to make up our minds about limiting our family, having another child, etc. He asked about Sally [who had been diagnosed as having the syndrome] and how she was.

That pleased me.'[20] But most counsellees arrive poorly informed.
They may need to learn some elementary genetics if they are to understand why only some members of their family have been affected or
why only some members are said to be at high risk of becoming
affected. Since the counsellor seems to know far more about these
matters than the client does, the counsellor is sometimes treated with
great deference. Deference to the counsellor's factual expertise rarely
causes problems. But many counsellors feel rather uncomfortable
when men and women defer to their moral judgements. And men
and women who have never really thought about the role of the
counsellor sometimes leave the clinic feeling slightly cheated.

> I tried to put him on the spot – what would you and your wife
> do? He wouldn't say definitely, but was encouraging. Said the
> risk was 10 per cent, something like that . . . then turned it
> around and said, 90 per cent okay.[21]

> I don't know how the medical profession expects us to react and
> behave in this situation because I don't want to go ahead and
> have another one and have everyone, the doctors basically,
> shaking their heads and saying, 'What a fool, doesn't she know
> this isn't done?'[22]

This kind of disappointment can be avoided by any counsellor who
is prepared to tell couples either that they should or should not try
again for a healthy child. But the directive counsellor exposes himself
or herself to quite devastating rejoinders.

> The doctor who delivered me said to have another baby. It
> annoyed me . . . because it could happen again, you can't tell
> someone to do it again.[23]

> The doctor told me not to have another baby. Said the risk was
> 50–50. You can't tell someone not to do it again. When you
> toss a coin nobody can tell you, even the cleverest doctor, which
> it is going to be – heads or tails – because it might be either.[24]

Some of the harshest lay critics of both 'directive' and 'non-
directive' genetic counselling remain unaware of the professional's
dilemmas until they counsel members of their own family. It is not
unknown for a vociferous opponent of directive counselling to advise

his own children or relatives against procreation; and the very men and women who express disappointment at a doctor's refusal to tell them that such-and-such a risk is acceptable often feel that it would be quite wrong for them to tell their own offspring that exactly the same level of risk should be run. Counselled parents who have produced one or more seriously affected children before any member of the medical profession told them that each procreative attempt carried a 1 in 2 risk of recurrence sometimes express resentment. If only they had been told earlier, they would not, they say, have taken the risk. But if a clinical genetics unit does ask them to take responsibility for warning their own children or relatives of the existence of exactly the same level of risk, the merits of early warning may lose their former obviousness. More accurately, perhaps, the rival merits of saying nothing, and letting (provident?) nature take its course, become more conspicuous.

Clinical geneticists sometimes use a coin to demonstrate the mathematical risks in autosomal dominant inheritance. Two coins will be needed if a heterozygous couple has difficulty grasping the quarter, half and quarter probability of the normal, heterozygous and homozygous states of autosomal recessive segregation. Where there's an empirical risk of 1 in 6, a die rather than a coin will be the most promising teaching aid.

Repeated throws or tosses of these familiar objects enable the genetic counsellor to drive home the message that 'chance has no memory'. The probability of heads turning up on the second toss of a fair coin is $\frac{1}{2}$, no matter what was the result on the first throw. Where two coins are being flipped together, the probability of getting two heads on any given toss remains $\frac{1}{4}$: these odds are as appropriate for the fifth joint flip as they were for the very first. And in the case of dice, the odds against getting a 3 on the eighth throw of a fair die with six faces are unaffected by the history of the first seven throws; the probability of a 3 on the eighth throw is $\frac{1}{6}$, as it was on the first throw.

In trivial situations where nothing of personal importance turns on the outcome of a toss of a coin or a roll of a die, children over the age of seven and adults of both sexes are likely to be extremely rewarding pupils for the mathematically-minded teacher. They quickly imitate the teacher's use of very simple numerical risk estimates and they soon acquire the habit of proclaiming that 'chance has no memory'. Psychologists have found, however, that much of this logical mathematical learning gets displaced or temporarily overridden when some individuals confront a situation in which one of the possible outcomes is either feared or strongly desired.

Whenever the result of a die throw really matters to us, we may behave as if the actions or wishes of the thrower could somehow override the familiar laws of probability. Not uncommonly, the thrower who needs a high number to win an exciting game throws very much harder than he or she throws when a small number is all that is needed. And many of us are prepared to stake far more money on the outcome of a future throw of a die than we will stake if a die has already been thrown but remains covered and out of sight, obscured by a handkerchief which cannot be twitched away until after the bet has been placed. Some of us may stop short of believing that our thoughts can scratch out the number on the uppermost face of a thrown die and then inscribe whatever number would win us the bet. But our rational scepticism is not boundless. It is sometimes very easy to believe that a prayer, an effort of will, or a 'special' throw, allows us to shorten the mathematical odds in respect of an uncertain outcome which has yet to be determined. This belief that our own words or gestures control events which have no known natural or logical connection with our verbal or non-verbal behaviour occurs in many contexts. It sometimes plays a role in reproductive decision-making under conditions of quantified uncertainty.

Another, quite distinct, belief is the belief that bad things happen to the doers of bad deeds whereas good things happen to the doers of good things; the universe is morally ordered and the distribution of suffering is not as random as it sometimes appears to be. Individuals who believe they are numbered among 'the good', or who believe they have done whatever is necessary to avoid being classified as 'bad', can expect to beat the mathematical odds. A zygote with genetically determined suffering in store for it will not have had very much time to call down suffering upon its own as yet unformed head by its own misdeeds, but in cases such as these the moral cause of its genotype may be sought in the behaviour of one or more of its parents and ancestors, or simply in God's mysterious will. When the past, present and future distribution of genetic suffering is viewed in this way, medical genetics does become a sort of Scientific Calvinism, reassuring to the elect but disturbing for the rest of us.

Genetic counsellors tend to avoid using coins or dice in cases where there's a multifactorial risk. If, for example, they want to convey the idea of a 1 in 35 risk of having a child with a particular disorder it is very much simpler to tell a couple that, 'If thirty-five families with exactly the same medical history come into the counsellor's office and then go home, and each of the couples produces one more child, thirty-four couples will have unaffected children, one couple will have an affected child, and no one can tell in advance which of these thirty-five children yours will be.' Formulations such as this are open to the objection that they exaggerate the representativeness of small samples (an *average* rate of one affected child per thirty-five does *not* guarantee that one affected child will appear in the next thirty-five births, or that two will appear in the next seventy births) yet such statements are more vivid for many men and women (and perhaps more useful) than a bare probability statement would have been.

Counsellors who are determined to give an unbiased account of the outlook for future offspring generally take care to avoid quoting a risk in only one way. If they say, for example, that a couple has a 1 in 10 chance of producing a child with a specific disorder they will be careful to allude to the fact that there's a 9 in 10 chance of an unaffected child. Similarly, in circumstances where there is a 1 in 4 chance of an affected child, reference will also be made to the 3 in 4 chance of producing a healthy one. Mathematically sophisticated readers who have never had occasion to visit a clinical genetics unit should not scoff at this spelling out of the complementary ratios. Any white-coated expert who talks only of possible misfortune – 'You've got a 25 per cent risk' – deserves to be regarded as biased.

Only rarely do we pass judgement on isolated tones or colours. In everyday life we perceive that one tone is louder than another; we note that a colour in one part of the landscape is darker than another. Some doctors and specialist genetic counsellors believe that express or implicit standards of comparison must be equally important in the perception of risk. In an influential paper published by the *British Medical Journal* almost a generation ago a doctor with extensive experience of counselling parents at the famous Great Ormond Street Hospital for Sick Children stated:

> The risk that any random pregnancy will end with some serious congenital malformation or other, or that some serious error will manifest itself in early life, is probably of the order of 1 in 40. This gives us, and our patients too, I think, a yardstick against which to measure the risk of recurrence of a particular condition. If a risk exceeds 1 in 10 then I think we can call it a bad risk; and probably most prospective parents would agree. If, on the other hand, the particular risk does not exceed, say, 1 in 20, we may call it a relatively good risk.[25]

More recently, another British medical geneticist has observed that:

> Many people do not have a clear idea of what constitutes a 'high' or 'low risk'. Thus some couples who are given a low risk (e.g. 1 in 200) express the view that this is far too high to be acceptable, whereas others . . . have been greatly relieved by a risk of 50 per cent. Clearly the nature of the disorder will determine what risk is acceptable, but it is helpful to be able to give some kind of reference point for comparison, such as the fact that 1 child in 30 in the population is born with a significant handicap, or that the population frequency of the disorder in question is (say) 1 in 200.[26]

An American clinical geneticist has warned that the exact phrasing of professional references to yardsticks matters. He recommends that couples be informed or reminded that each pregnancy carries a 3–4 per cent risk of a child being born with a handicapping condition. The risk figure quoted for their particular disorder will be over and above this level of risk. In the case of a couple who have previously had a child with a unilateral cleft lip (and who have no similarly affected family relatives) the risk of producing another child with the same disorder will be 3 per cent. Had this couple been counselled before the arrival of their affected first-born the risk of having a child with a birth defect would have been the general population risk of

3–4 per cent. But now, taking into account their production of one affected child, the risk is 6 or 7 per cent (3 + 3 per cent or 3 + 4 per cent). Any counsellor who contents himself or herself with saying, 'Your chances of having a child with a birth defect are about twice as high as that of other couples', will convey a very different message from the one that would be conveyed if he or she incorporated the same yardstick but focused instead on the likelihood of an *un*affected child: 'Before the birth of your child with a cleft lip, the likelihood of your having a normal child was 97 or 96 per cent; now it is 94 per cent.'[27]

Some counsellors comment on the risk figures they give. 'The chances of your potential grandchildren marrying a husband with the same gene are *extraordinarily small*, of the order of one in tens of thousands. I think, therefore, that you can rest assured that they run *an extremely small risk* of either contracting the illness themselves or passing it on to their children.'[28] In one study of counselling sessions at a children's hospital in Canada, counsellors frequently used descriptive tags when the risk was less than 10 per cent but never when it was any higher.[29]

Recipients of risk information sometimes comment immediately on figures quoted to them. Some men and women express relief at the 'good' news. This reaction is not confined to prospective parents who face a recurrence rate lower than 1 in 20; risks as high as 1 in 2 have been greeted in exactly the same way. In time, however, many counsellees re-examine the figures and some counsellees are then disappointed to discover that although rate information limits their uncertainty, it does not banish uncertainty altogether. Contemplation of a 1 in 1,000 chance of something going wrong, no less than contemplation of a 1 in 2 chance, directs attention to the possibility of misfortune. Both levels of risk are likely to be converted into exactly the same binary form: '*Either* our child will be affected *or* our child will be normal.'

Not knowing for certain whether or not a future child would be affected, some counsellees make sustained or fleeting efforts to imagine what would be involved. A shift in focus from risk figure to outcome may be normal in female counsellees. This may also be true of their male partners.

Some students of reproductive behaviour have assumed that each genetic condition carries a quite distinctive physical, emotional, financial and social cost or burden. Other researchers doubt the assumption's validity, and these more cautious investigators are beginning to pay very close attention to the subjective perceptions of the men and women who actually make the reproductive decisions.

A group of fifty or so female counsellees who had been referred to the Montreal Children's Hospital in Canada spoke of the repercussions of a genetic disorder in a much broader way than many doctors would expect. For many of these women, genetic counselling was not the soon-to-be-forgotten prelude to an almost instantaneous decision. Important ambiguities made it impossible in many cases to be sure about the extent of future handicap or ill health. Science might provide an account of the range of severity but it was usually impossible for the geneticist to guarantee that a future affected child would have exactly the same symptoms as patient *A* or patient *B*. Knowing the name of a threatening genetic condition was not enough. Nor was it sufficient to know the range of symptoms or even the average symptoms. Counsellees deliberating the morality and the desirability of another child wanted to know what an affected child would be like.

Some of the individuals who learn of their risk status after the diagnosis of a genetic disorder in one or more of their children yearn for the innocent years when family planning was a relatively casual matter. Confronted all of a sudden with the need to consider for themselves the value of having children, and knowing that their decision is being scrutinised by a professional expert, genetic counsellees may feel cursed, not only by the genotype or medical history which puts future children at risk but also by the need to decide whether to take the risk. Even when a decision has been made, some counsellees will review their choice, and wonder whether it was truly the right one.

The decision-making process itself may be characterised by simplifying strategies. It has been suggested that probabilities are relegated to a very minor role once they have alerted an individual or couple to the possibility of misfortune in their family. Parents and prospective parents shift their attention from specific risk figures and concentrate instead on scenarios constructed to test their ability to cope if the worst or something approaching it were to happen.

Although many people spontaneously generate their own scenarios, some genetic counsellors use prompt sheets. An individual or couple might be handed a form (adapted to reflect the counsellor's understanding of the relevant facts) which invites the counsellee(s) to imagine that:[30]

> Your child would always need the same care as an infant needs
>
>> How do you feel you would respond?
>> Which of the following statements most closely reflects your feelings?
>>
>>> (a)   I/we could cope
>>> (b)   I/we don't know whether I/we could cope
>>> (c)   I/we couldn't cope
>>> (d)   Other – please explain
>
> Your child's condition prevented her/him from attending ordinary school and pursuing a career
>
>> How do you feel you would respond?
>> Which of the following statements most closely reflects your feelings?
>>
>>> (a)   I/we could cope
>>> (b)   I/we don't know whether I/we could cope
>>> (c)   I/we couldn't cope
>>> (d)   Other – please explain
>
> Your child's condition required special equipment, and expensive hospital treatment
>
>> How do you feel you would respond?
>> Which of the following statements most clearly reflects your feelings?
>>
>>> (a)   I/we could cope
>>> (b)   I/we don't know whether I/we could cope
>>> (c)   I/we couldn't cope
>>> (d)   Other – please explain
>
> You have a healthy child, however she/he may be a carrier for a genetic disease
>
>> How do you feel you would respond?

Which of the following statements most clearly reflects your feelings?

- (a)   I/we could cope
- (b)   I/we don't know whether I/we could cope
- (c)   I/we couldn't cope
- (d)   Other – please explain

You have decided not to have a child that is genetically yours

How do you feel you would respond?

Which of the following statements most clearly reflects your feelings?

- (a)   I/we could cope
- (b)   I/we don't know whether I/we could cope
- (c)   I/we couldn't cope
- (d)   Other – please explain

The absence, at the moment, of cheap and reliable means of screening entire populations for every harmful gene under the sun means that much genetic counselling is done on a retrospective basis. Very often it is the arrival of an affected child that first alerts a couple's doctor to the existence of specific risks for subsequent children. As a class, women who have produced at least one normal child in addition to their affected child may be more confident of their ability to produce future normal children than are women who have no healthy children to their credit. Having had powerful evidence that the worst need not happen, members of the former class are readier, perhaps, to believe that being at risk can be dealt with.

Many counsellees are very confident of their own ability to process the information and to act upon it in the best way. They relish the sense of being in sole or joint control. But some men and women seem to hope that somebody else will take the decision for them. Reluctant decision-makers may repeatedly ask the counsellor what the counsellor and the counsellor's spouse would do. Or they may procrastinate under a cloak of market research – quizzing all their relatives and friends to establish the likely level of approval for various courses of action or inaction.

Reproductive roulette may be played by the person who wants to diffuse responsibility for future events. Almost half the women in a small group of female counsellees interviewed some time after counselling said that they had used contraceptive practices which were known by them to be insufficient for guaranteeing that a pregnancy would not occur. None was absolutely confident at the time of her ability to cope with an affected child. None had the option of prenatal diagnosis. Of those who did become pregnant, many were relieved. The 'accident' of another pregnancy had put an end to the waiting and saved the mothers from becoming too frightened to produce a child. Individuals who leave the door to reproduction ajar in this way, so that Fate has a good chance of bringing a pregnancy which is partly wanted and partly feared, are not necessarily indecisive. Some are plumping for the course of action which seems most likely to neutralise their dilemma.

Prenatal tests

Diagnostic tests performed after conception but before a pregnancy has gone to term are regarded by some women as a blessing which liberates them from the need to devote nine months and sometimes many years to a creation that has gone wrong. Other women and couples and doctors are ambivalent or even hostile.

> A couple of years ago [1981] I became pregnant. The amniocentesis test showed that the child had Down's syndrome. Life is full of surprises, but I can't say that this was one of them. Similar things had happened to too many of my friends for me to imagine that 'it can't happen to me'. So when the card from the hospital arrived summoning me to an unexplained early appointment, I was in no doubt as to the probable reason.
>
> The episode, however, was not lacking in eye-opening moments. The first of these had occurred some weeks previously when my assumption that, being in my late thirties, an amniocentesis test would be performed as a matter of course, was rudely shattered by my GP. 'If you can find anyone to do it for you,' he said, in a tone that indicated that he personally (a) disapproved and (b) could see no reason for the fuss.
>
> I went to the hospital geared up for a fight, only to find that, for anyone over 35, it was indeed a matter of course. Anyone

else who requests the test can also have one at this hospital, which is particularly good on maternity services, but this is by no means universal practice elsewhere. An acquaintance who was not yet 35, but who nevertheless very much wanted the test, was refused one in London.

As it turns out, her child was all right; and as it turns out, I was able to have the test. But how would we have felt if I had not had the test, or if her child had been born with a prenatally detectable defect?

My next surprise came when I discussed the bad news with friends. Naturally – or naturally is how it seemed to me – I wanted an abortion. But judging by the number of people who told me I had 'made the right decision' there are those who would not see this as the inevitable course. My decision was made when I decided to have the test: why have it, otherwise? Yet I now know that there are those who, having been told that the baby will be defective, still elect to have it.

This I cannot understand, although in such a situation clearly everyone must do as she thinks fit. What I cannot tolerate is the thought that there are people who would have wished to prevent me acting as I did. When I look at the normal, healthy baby daughter whom I had sixteen months later, and think that, instead of her, these people would, if they could, have condemned me to have an avoidably handicapped child, my blood runs cold.

I elected to have an abortion. What I had in mind was a swift routine operation during which I should be mercifully unconscious. But the amniocentesis test cannot be performed before 16 weeks – not enough amniotic fluid – and the results are not available for at least two weeks, often longer. By this time it is too late for a surgical abortion and what must be undergone is an induced birth.

Any merely physical ordeal, though, is as nothing compared to the psychological trauma of this situation. Clearly, it is very difficult for hospital staff, or indeed anyone, to gauge the best psychological approach to adopt towards the ex-mother. I was treated with enormous kindness and sympathy – something which is by no means true for most routine abortions – and for this I was deeply grateful. But what struck me was that everyone expected me to go to pieces. 'You can expect to have quite a reaction after this,' said one doctor. The obstetrician repeatedly told me how well I was taking it. Friends have since asked me, in awestruck tones, how I could bear to undergo it all.

Kindness is clearly essential. But I wonder how many women, offered this open invitation to repine, don't thankfully grasp it simply because this is what is expected? After all, we tend to fulfil unspoken expectations.

Yet this situation, though sad, is not terminally depressing in the way that having the child undoubtedly would have been. Indeed, because one is having the abortion for such an excellent reason, it is probably less depressing than many abortions performed for other reasons.

The essential thing, after all, is that, for most of us, there is always another chance. Which brings me to my final shock. When, a month later, I saw the obstetrician for a check-up, he asked me about my future plans. I told him we thought we would have one more try, and if anything went wrong this time, that would be that. 'In your case', he said, 'I think that's probably a very good idea.' Slightly startled, I asked him if there were any cases in which he would not recommend trying again? He assured me that there were quite a number.

I left wondering how I should feel had I been advised not to try again. Presumably, the judgement is based on how the woman takes the initial disappointment. Statistically, the odds against conceiving another Down's syndrome child are much shorter in the case of a woman who has already shown this predisposition. But, in fact, one would have to be very unlucky to have this happen twice; and I can't help feeling that the main thing which kept me on a relatively even keel was the same thing that had stopped me having a baby earlier – namely, having a full and interesting life which would be curtailed, at the very least, by children. Surely a more violent reaction might indicate a stronger need to have a baby?

Of course, advice is there to be ignored. But most women consulting a doctor about these matters are not hardened cynics. And even I – tougher than most – worried about how I should face my doctor if I disregarded his advice, which was not based on any real knowledge of myself or my affairs, and subsequently needed his help.[31]

Another woman accepted that an amniocentesis test could accurately identify fetuses with Down's syndrome but, having already given birth to a child with the syndrome, six years earlier, she had reservations about selective abortion. 'If I hadn't already had one it would be an easier decision but I've had W— and she is classed as handicapped. But she's lovely, she seems as normal as can be, so I couldn't have an abortion after W—, but if I hadn't had her, my idea of *being* handicapped would be different.'[32]

When a 39-year-old practising Catholic became pregnant for the first time and was advised of the availability of amniocentesis, she was unable to avoid a decision about amniocentesis for Down's syndrome and other abnormalities. 'My husband and I decided to chance it in the end and take whatever God sent.' She gave birth, just short of her fortieth birthday, to a perfect baby girl.[33]

A 36-year-old woman became pregnant during her second marriage, thirteen years after having her last baby. She could have persuaded her doctor to allow her to have amniocentesis on the National Health Service but he said that in her area it wasn't done until women had entered their forties. She did not insist on a test, and gave birth to a child with Down's syndrome who died ten days later: 'We decided not to go ahead with vital operations. When very soon afterwards I was pregnant again there was no question of not having the test. It was done at 16 weeks and I knew at 19 weeks that the child – now 3-year-old Polly – was normal.'[33]

GOOD LUCK!

Notes

1 Introduction

1. 'Should we leave the fruits . . . raw nature?' Joseph Fletcher, *The Ethics of Genetic Control, Ending Reproductive Roulette* (New York, 1974), p. 36.

2. In everyday conversation the word 'risk' has many meanings. It sometimes means the *probability* of an undesirable event. Sometimes it refers to *events* themselves – the 'risks' of air travel, for example. If we consider not only the implications of air-sickness (or the implications of a mid-air collision) but also the probability of such an event we may then be in a position to make an informed statement about the *seriousness* of the risk. The timorous holiday-maker may refuse to take a 1 in 1 million chance of death in mid-air. A more sanguine traveller may face such a risk with equanimity, but still avoid air-travel because the risk of being grounded at some dreary airport seems unacceptable.

   Medical geneticists have a habit of classifying risks greater than 1 in 10 as 'high' risks. See Chapter 8.

3. 'For in the case of what other diseases . . . a Down's.' Leon R. Kass, 'Implications of prenatal diagnosis for the human right to life' in Bruce Hilton and Daniel Callahan (eds), *Ethical Issues in Human Genetics* (London, 1973), p. 190.

4. In this book 'embryo' is used in relation to the first stages after fertilisation. 'Fetus' is reserved for later stages. The Warnock Committee of Inquiry into Human Fertilisation and Embryology chose to regard the

embryonic stage as the six weeks immediately following fertilisation.

5. Lawyers will know that the Congenital Disabilities (Civil Liability) Act applies in England, Wales and Northern Ireland. Non-lawyers should note that the Act does not extend to Scotland. In this context, as in many others in this book, English law has limited application.

6. 'The HOMUNCULUS, Sir, ... state and relation.' Laurence Sterne, *Tristram Shandy* (London, 1760), vol. I, chap. II.

7. 'In general parlance ... God's will.' Nancy Wexler, 'Genetic "Russian roulette": the experience of being "at risk" for Huntington's disease' in Seymour Kessler (ed.), *Genetic Counseling: Psychological Dimensions* (London and New York, 1979), pp. 209-10.

8. 'That which we call fortune ... God.' Thomas Cooper, *Certaine Sermons* (London, 1580), p. 164.

9. 'Building countless universes ... inexplicable phenomena.' Erasmus, *Praise of Folly*, translated by Betty Radice (London, 1971), p. 151.

10. 'Whereas many persons ... they are in.' John Graunt, *Natural and Political Observations mentioned in a following Index, and made upon the Bills of Mortality* (London, 1662), chap. II, para. 17.

11. 'It is ... to avoid it.' The Port Royal, 'Art of thinking' in *Course of Education Pursued at the Universities of Cambridge and Oxford* (London, 1818), vol. 3, Part IV, chap. XVI.

12. Founded in the seventeenth century, the Royal Society of London received a charter from Charles II. It was hoped that the Society would improve natural knowledge.

13. 'It is odds ... to the females.' John Arbuthnot, *Of the Laws of Chance* (London, 1692), preface.

14. 'To judge the wisdom ... constant proportion.' John Arbuthnot, *Philosophical Transactions of the Royal Society of London* (1710), vol. 27, p. 188.

15. 'What can the maintaining ... or of any other creature ...!' William Derham, *Physico-Theology* (London, 1713), Book IV, chap. X, p. 178.

16. Toward the end of Anne's reign, shortly before the Hanoverian succession, Derham observed that 'Major Graunt, (whose conclusions seem to be well-grounded,) and Mr King, disagree in the proportions they assign to males and females. This latter makes in London, 10 males to be to 13 females; in other cities and market towns, 8 to 9; and in the villages

and hamlets 100 males to 99 females. But Major Graunt, both from the London and country bills saith, there are 14 males to 13 females . . .

This proportion of 14 to 13, I imagine is nearly just, it being agreeable to the bills I have met with, as well as those in Mr Graunt. In the 100 years, for example, of my own parish register, although the burials of males and females were nearly equal, being 633 males, and 623 females in all that time; yet there were baptised 709 males, and but 675 females, which is 13 females to 13′7 males.'

In 1962 Cedric Carter noted that England and Wales were yielding about 106 boys born alive for every 100 girls.

In 1980 James S. Thompson and Margaret W. Thompson noted that in North America the sex ratio at birth was about 1.05 (105 boys for every 100 girls).

17.   'The old argument . . . fails . . . wind blows.' Charles Darwin, 'Auto-biography' (1876) quoted in Francis Darwin's *The Life and Letters of Charles Darwin* (London, 1887), vol. 1, p. 309.

18.   'So soon as it becomes common knowledge . . . profoundly changed.' W. Bateson, *The Methods and Scope of Genetics* (Cambridge, 1908), pp. 34–5.

19.   'Sir . . . Scientific Calvinism.' Beatrice Bateson, *William Bateson FRS, Naturalist* (Cambridge, 1928), p. vi.

20.   'All the ordinary . . . composition.' W. Bateson, *The Methods and Scope of Genetics* (Cambridge, 1908), pp. 4–6.

21.   'To get a true picture . . . ingredients in our list.' Ibid., pp. 7–8.

22.   '[Until Mendel began his analysis] the existence . . . 4,000 years.' Quoted by Beatrice Bateson, *William Bateson FRS, Naturalist* (Cambridge, 1928), p. 279, being part of William's 1914 Presidential Address to the British Association for the Advancement of Science (Australian Meet-ing, Melbourne).

23.   'If, instead . . . social usage.' Quoted by Beatrice Bateson (above) at pp. 299 and 303. This Presidential Address was prepared for the Sydney Meeting of the BAAS.

24.   This book makes no attempt to expound such matters as the normal genetic control of protein structure and synthesis. Nor does it expound experimental molecular hybridisation, DNA fractionation, gene-mapping, cloning and the preparation of gene libraries. D. J. Weatherall's

The New Genetics and Clinical Practice, published late in 1982, provides lucid detail on these and other matters. See also Alan E. H. Emery's *An Introduction to Recombinant DNA* (1984).

2   Genes, Chromosomes and Patterns of Inheritance

1. 'The linkage . . . is cellular.' Clifford Grobstein, *From Chance to Purpose* (Reading, Mass., 1981), p. 80.

2. This part of the book is about biology. Although some biologists wince when they hear talk of life starting at any point in a cycle it is not wildly unorthodox to say that 'we' started life as a zygote. The statement refers to the zygote's totipotentiality – from the zygote there developed the tissues and organs that make up a human body (plus tissues destined to form parts of the afterbirth).

3. 'Every child . . . as the actual ones.' François Jacob, *The Possible and the Actual* (London, 1982), p. 8.

4. M. A. Patton, Honorary Senior Registrar in Clinical Genetics at the Great Ormond Street Hospital for Sick Children in London comments: 'From reports in the press, mutations are caused by radiation and chemicals but it seems that most germinal mutations are not caused in this way but they are random copying errors. I often explain this to couples by analogy. We have 50,000 structural genes and we have to make a copy of each one to make each egg or sperm. Occasionally things go wrong. In a similar way if we were asked to make 50,000 photocopies, every now and again, one would print with a smudge or pass a blank sheet through without printing. The copying of genes is very much more efficient but very occasionally an error will be made by chance alone.'

5. There is an alternative explanation. Miss Brown may have just one parent who is a heterozygote; a mutation may have arisen in the gamete from the other parent. In many autosomal recessive disorders such an event will be hundreds of times less likely than a mating between two heterozygotes. It seems likely, therefore, that both Mr and Mrs Brown carry the mutant gene.

    In the past, the 'invisibility' of recessive genes necessitated inference. Today long-standing biochemical techniques and recombinant DNA methods are making direct examination of genes the norm. Clinical geneticists have a growing battery of gene detection tests.

6.   Homozygous females in X-linked disorders are extremely rare in the serious genetic disorders.

7..   This statement is usually true. But it is important to note that fresh germinal mutations are not unknown in X-linked inheritance.

8.   One of Queen Victoria's biographers has stated that a third daughter, the Princess Royal, carried the haemophilia gene. For divergent views on a very famous medical pedigree see Elizabeth Longford's *Victoria R I* (London, 1964); Victor A. McKusick's *Human Genetics* (2nd edn, Englewood-Cliffs, New Jersey, 1969); and the British Museum's *Origin of Species* (Cambridge and London, 1981).

9.   'A cloud of worry . . . come from?' Elizabeth Longford, *Victoria R I*, p. 235.

10.   At birth all the oocytes are present in the ovary and they are released one by one at each ovulation. Some scientists would expect the health and genetic integrity of these gametes to decline with maternal age. By contrast, spermatogenesis is a continuous process – new batches of sperms are produced all the time, with less chance of genetic 'decay'.

11.   'It's just been an accident . . . our family.' Adapted from Patricia T. Kelly, *Dealing with Dilemma* (New York, 1977), p. 32.

12.   James S. Thompson and Margaret W. Thompson observe, at p. 262 of *Genetics in Medicine* (Philadelphia, 1980), that: 'The multifactorial/ threshold model has not been accepted without criticism. In brief summary, the criticisms are based on the difficulty of testing some of the underlying assumptions and the fact that the data will fit other models, especially if heterogeneity or reduced penetrance is invoked. However, multifactorial inheritance does explain many of the formerly puzzling features of the family distribution of a number of disorders, and has stimulated research into the nature of the underlying factors that contribute to their causation.'

13.   Readers who have never given much thought to probability may be no worse off than some specialists in that vast and complex field.
     For a sight of some of probability's many facets see A. J. Ayer's *Probability and Evidence*; the three works listed under John Cohen's name in the bibliography near the end of this book; A. W. F. Edwards's *Likelihood*; Richard Eggleston's *Evidence, Proof and Probability*, D. A. Gillies's *An Objective Theory of Probability*; Ian Hacking's *The Emergence*

of *Probability*; Darrell Huff's *How to Take a Chance*; D. V. Langley's *Making Decisions*; Russell Langley's *Practical Statistics: Simply Explained*; L. E. Maistrov's *Probability Theory: A Historical Sketch*; Peter G. Moore's *Reason By Numbers*; F. Mosteller *et al.*'s *Probability with Statistical Applications*; Edmund A. Murphy's *Probability in Medicine* and the same author's paper 'How much difference does the use of Bayesian probability make?' in *Counseling in Genetics*, edited by Y. E. Hsia *et al.*; and Karl R. Popper's *Realism and the Aim of Science*. (Years of publication are given in the bibliography, which also lists additional works on the subject.)

3 Conception and Development

1. 'Of Bodies, and of Mans Soul.' Part of the title of a work by Sir Kenelm Digby (London, 1669).

2. It is impossible to be very precise about the timing of many of the important events in embryology. Scientists have provided us, however, with a general schedule.

3. 'By the ninth day ... toes and ears.' Abigail Lewis, *An Interesting Condition* (London, 1951), pp. 38–9.

4. Marjorie A. England's *A Colour Atlas of Life Before Birth* (London, 1983) has endoderm forming at about the seventh day after fertilisation of the human egg; ectoderm and mesoderm are formed from the remaining cells of the inner cell mass during week 3.

5. Marjorie England (in common with Alfred Sherwood Romer and Thomas S. Parsons, authors of *The Vertebrate Body* (London, 1977)) distinguishes between the body or surface ectoderm and neural ectoderm. Readers who are curious about the topographical origins of the specialised cells in their bodies will turn, no doubt, to the works cited above and to B. I. Balinsky's and B. C. Fabian's *An Introduction to Embryology* (Philadelphia, 1981). The diary, a highly skilful literary work, was never intended as a science manual.

6. Balinsky and Fabian note at p. 574 that 'differentiation' has at least two meanings. In its broader sense differentiation is 'the process in which the cells or other parts of an organism become different from one another and also different from their previous condition'. Used more narrowly the term refers to the process which results in cells and groups of cells acquiring the ability to perform their special functions.

The embryonic cells and tissues usually follow a consistent pattern of movement and arrangement. Experiment suggests, however, that what any given cell becomes depends not on its position a few days after fertilisation but rather on the potentialities in it and on the influences it is subjected to.

According to von Baer's 'law' the features that are characteristic of all vertebrates (segmented muscles, brain and spinal cord, for example) develop earlier, in the embryo, than the features that distinguish one class of vertebrates from another (hair in mammals, feathers in birds, and so on).

7. Marjorie England notes that at about day 28 the embryo is flexed in a 'C' shape. Later on, it uncurls to some extent. In England's schedule the forebrain divides into three vesicles in week 5, the midbrain remains as before but the hindbrain forms two regions.

   The 'gill clefts' are never gills.

8. The embryonic and fetal membranes and the placenta not only protect the embryo or fetus; they also allow nutrition, excretion and respiration during development. England notes that the first vascular blood forms from yolk sac mesoderm at about day 13. This will provide embryonic blood until the liver starts to form blood about three weeks later. It also provides nutrients during weeks 2 and 3 while the chorioallantoic placenta is developing. The early embryo has an umbilical stalk containing one large umbilical vein, two umbilical arteries and the allantois. The arteries transport blood from the embryo to the chorionic villi and the umbilical vein returns blood to the embryo.

   By six weeks the body length of the embryo or fetus, as measured from crown to rump, may be about 20 mm.

9. 'Births always consist of two-fold seed . . . more than equal share.' Titus Lucretius Carus, *On the Nature of Things* translated by Thomas Jackson (Oxford, 1929), p. 152.

10. 'The perfect blood . . . vine distills.' *The Divine Comedy of Dante Alighieri* translated by Henry Wadsworth Longfellow (London, 1890): Purgatory, Canto XXV.

11. For centuries theologians and canonists held that eighty days passed between conception and the animation of a female fetus. Forty days sufficed for a male fetus. (The Book of Leviticus discriminates to the same extent, though in a different context – eighty days of postnatal purification are prescribed for the mother of a girl whilst forty days'

purification suffice for the mother of a boy.) Aristotle, in Richard Creswell's 1862 edition of the *History of Animals*, had a slightly different schedule. Males were animated at about forty days after conception but females were not animated until about the ninetieth day.

12. 'A thinking, intelligent being . . . times and places.' John Locke, *Essay Concerning Humane Understanding* (London, 1694), Book II, chap. XXVII, para. 9.

13. 'Respect for human life . . . its own growth.' 'Declaratio de abortu procurato' in *Acta Apostolicae Sedis*, vol. 66 (1974) translated and cited by W. A. R. Thomson, *A Dictionary of Medical Ethics and Practice* (Bristol, 1977), p. 1.

14. 'Even supposing a belated animation . . . his soul.' (1974) *Acta Apostolicae Sedis* at p. 738 (Latin) cited in translation by Catholic Bishops' Joint Committee on Bio-ethical Issues on behalf of the Catholic Bishops of Great Britain, *In Vitro Fertilisation: Morality and Public Policy* (Abbots Langley, 1983), pp. 7–8.

15. Some would regard a total of 15–20 per cent natural loss as a more realistic figure – see *Human Concern*, No. 18, Spring 1985, at p. 3. *Human Concern* is published by the London-based Society for the Protection of Unborn Children.

16. 'Tonight, lying in bed . . . rather awesome.' Abigail Lewis, *An Interesting Condition* (London, 1951), pp. 46–7.

4 Prenatal Tests

1. 'The once seemingly . . . now is upon us.' Thaddeus E. Kelly, *Clinical Genetics and Genetic Counseling* (Chicago and London, 1980), p. 340.

2. See the prologue to the 'The Wife of Bath's Tale' (line 677) in *The Canterbury Tales*, edited by A. C. Cawley (London, 1975).

3. '. . . describe the nature of the womb . . . and another.' *The Notebooks of Leonardo da Vinci*, edited and translated by Edward MacCurdy (London, 1956), vol. I, p. 126.

4. In 1838, when Francis Galton was an indoor pupil in the Birmingham General Hospital, at Snow Hill, 'The stethoscope was considered generally to be new-fangled; the older and naturally somewhat deaf practitioners pooh-poohed and never used it.' Galton could never understand

why 'Youths selected for their powers of sharp hearing should not be so far instructed as to be used by physicians, much as pointers and setters are used by sportsmen. They could be taught what to listen for, probably by means of some sound-emitting instruments more or less muffled, and how to describe what they heard.' Sir Francis Galton, *Memories of My Life* (London, 1908), p. 29.

5. 'I felt a couple . . . itself.' Sheila Kitzinger, *Birth Over Thirty* (London, 1982), p. 43.

6. 'The needle in my stomach felt so threatening.' Ibid., p. 42.

7. 'For all I know . . . lamb.' Ibid., p. 45.

8. 'It was fascinating . . . decision.' Ibid., p. 35.

9. Maternal blood sampling to check protein levels and a number of other factors is routine in some parts of the world. But the blood samples are not yet used to establish fetal cell lines for detailed analysis.

10. 'The once seemingly science fiction idea . . . prenatal diagnosis.' Thaddeus E. Kelly, *Clinical Genetics and Genetic Counseling* (Chicago and London, 1980), pp. 340–41.

5   Abortion

1. 'As conception . . . abortion and sterilisation.' Jane E. Hodgson, 'Conclusion' in Jane E. Hodgson (ed.), *Abortion and Sterilization: Medical and Social Aspects* (London, 1981), p. 545.

2. 'As to exposing . . . sensation and being alive.' Aristotle, *The Politics* translated by H. Rackham (London, 1932), Bk. VII: 14.

3. The principle of extraterritoriality gave certain people exemption from the jurisdiction of local laws and tribunals.

4. 'It is permissible . . . unformed in human shape.' Isam R. Nazer in Robert E. Hall (ed.), *Abortion in a Changing World* (London, 1970), vol. I, p. 268.

5. Abandoned in this context, the distinction between animated and unanimated fetuses nevertheless continued to be important with regard to vindictive penalties, and especially irregularity (P. Charles Augustine, *A Commentary on the New Code of Canon Law* (London, 1920 and 1922), vols IV and VIII respectively).
    Readers who find themselves irritated by the obvious limitations of

this chapter's compressed account of vivified, animated, formed and ensouled organisms will be soothed, perhaps, by Roger John Huser's *The Crime of Abortion in Canon Law* (Washington, D C, 1942); by Glanville Williams's *The Sanctity of Life and the Criminal Law* (London, 1958); by Germain Grisez's *Abortion: The Myths, the Realities and the Arguments* (New York, 1972); by John Thomas Noonan Jr's (ed.), *The Morality of Abortion* (Cambridge, Mass., 1970); by Joseph F. Doncel's 'Immediate Animation and Delayed Hominization' in *Theological Studies*, vol. 31 (1970), pp. 76–105; by the historical appendix to the *Report of the (Lane) Committee on the Working of the Abortion Act* (London, 1974, Cmnd 5579); and by John Mahoney's *Bioethics and Belief* (London, 1984).

6. By virtue of the Criminal Law Act, 1967 offences are no longer classified as misdemeanours and felonies.

7. See pp. 81 3.

8. 'For the stage prior . . . life or health of the mother.' (1973) 410 United States Reports (Washington D C), p 164

9. 'Every woman, being with child . . . like intent.' *Halsbury's Statutes of England* (3rd edn, London, 1969), vol. 8, p. 168.

10. Originally section 58 referred to 'a felony', whilst section 59 referred to 'a misdemeanour'. As amended by the Criminal Law Act, 1967, both sections create 'an offence'.

11. 'Any person who . . . child destruction.' *Halsbury's Statutes of England* (3rd edn, London, 1969), vol. 8, p. 304.

12. Originally section 1(1) of the 1929 Act created 'a felony'. As amended, section 1(1) creates 'an offence'.

13. The Act provides that no person shall be under a duty to participate in any treatment under the Act to which he or she has a conscientious objection. But this does not affect a person's duty to participate in treatment necessary either to prevent grave permanent injury to the physical or mental health of a pregnant woman or to save her life.

14. 'There is . . . protecting someone else.' Linda Francke, *The Ambivalence of Abortion* (London, 1979), pp. 17 and 7.

15. 'Emotionally there is almost always . . . fooling themselves.' Ibid., pp. 18 and 11.

16. Thirteen families is clearly a very small sample. The study by Bruce D.

Blumberg, *et al.* (see bibliography for details) is valuable because it sheds light on individual responses.

6   Contraception and Sterilisation

1.   'A contraceptive differs . . . been conceived.' Soranus' *Gynecology* translated by Owsei Temkin (Baltimore, 1956), p. 62.

2.   'Just as every season . . . during intercourse.' Ibid., p. 34.

3.   Progestogen is synthetic progesterone.

4.   'The direct interruption . . . end or as a means.' *Humanae Vitae* (1968) translated by Claudia Carlen, *The Papal Encyclicals 1958–1981* (Washington, DC, 1981), pp. 226–7.

5.   'Unconditional respect for nature . . . human responsibility.' Hans Küng, *Infallible?* (London, 1971), p. 31.

7   Artificial Reproduction

1.   'Be it enacted . . . *in vitro* fertilisation.' Enoch Powell's unsuccessful *Unborn Children (Protection) Bill* (London, 1984), preamble and clause 1.

2.   'The idea that adultery . . . occasion to wrestle.' [1958] Session Cases (Edinburgh, 1958), p. 108.

3.   '. . . enquire into . . . legal consequences', 'taking account . . . desirable.' *Report of the Departmental Committee on Human Artificial Insemination* (London, 1960, Cmnd 1105), p. 1.

4.   Confidentiality may be more difficult to guarantee in the National Health Service than in a private clinic.

5.   For extra detail and comment see R. Snowden's and G. D. Mitchell's *The Artificial Family* (London, 1981) and Gillian Hanscombe's 'The right to lesbian parenthood' in *Journal of Medical Ethics*, vol. 9 (1983). This list and the list that follows it owe much to Snowden's and Mitchell's work.

6.   By virtue of the Perjury Act, 1911 the wilful making of a false statement to the registrar of the child's birth in order to procure the making of an erroneous entry in the register is an offence.

7.   'The fidelity of marriage . . . distinguished from love.' Joseph Fletcher, *Morals and Medicine* (London, 1955), p. 139.

8. 'Is artificial fecundation of women permissible? Non licere.' Henry Davis, *Artificial Human Fecundation* (London, 1951), p. 12.

9. 'There is no bodily impairment ... aesthetically significant.' Social Welfare Commission of the Catholic Bishops' Conference (England and Wales) (Abbots Langley, 1983), p. 24.

10. 'A further concern ... is recalled.' Ibid., p. 31.

11. In the United Kingdom and in many parts of the world the demand for children to adopt exceeds supply. Adoption agencies are highly selective.

    Men and women who decide to adopt in order to bypass a genetic risk sometimes experience exceptional difficulties. Their own life expectancy or state of health may push them to near the bottom of agency waiting-lists. Some agency workers seem to be prejudiced against applicants who are capable of creating a child but who elect not to do so. Determined applicants will apply early, to more than one agency if necessary.

12. Depriving the donor of parental rights and duties would have the effect, in the context of AID provided to an unmarried woman, of blinding the eyes of the law: the AID child would have no father.

13. 'He felt the child ... disappeared.' Naomi Pfeffer and Anne Woollett, *The Experience of Infertility* (London, 1983), pp. 71–2.

14. 'My ex-wife and I ... my son.' R. Snowden and G. D. Mitchell, *The Artificial Family* (London, 1981), p. 29.

15. 'Any child we have ... my wife.' Ibid., p. 29.

16. Early this century researchers stomped America in an ambitious attempt to work out the genetics of 'sublimity' and 'stolidity'.

17. 'We see the child as ours.' R. Snowden and G. D. Mitchell, *The Artificial Family* (London, 1981), p. 36.

18. The Committee did not suggest that this practice was universal in America. It was a feature of some of the clinics there.

19. 'AID regards semen ... nature of man.' David Ison, *Artificial Insemination by Donor* (London, 1983), p. 16.

20. 'There is nothing that we would change ... well-balanced adults.'

R. Snowden and G. D. Mitchell, *The Artificial Family* (London, 1981), p. 76.

21. 'My mother conceived me . . . husband in any way.' Ibid., pp. 88–90.

8   Decisions

1. 'When you toss a coin . . . might be either.' Anonymous 12-year-old girl cited by John Cohen, *Chance, Skill and Luck* (London, 1960), p. 36. The original quote has been modified for the purposes of this chapter.

2. Blood grouping tests and increasingly sophisticated biochemical tests can demonstrate, or suggest, that an individual is not the biological father of a particular child.

    Following egg donation, H L A-typing and other tests may be needed to establish maternity. (Human leucocyte antigens are inherited substances on the surface of the cells throughout our body.)

    See Alec J. Jeffreys, John F. Y. Brookfield and Robert Semeonoff, 'Positive identification of an immigration test-case using human D N A fingerprints' in *Nature*, vol. 317 (1985), pp. 818–19.

3. 'To lessen anxiety . . . can be useful.' Joseph Adams, *A Treatise on the Supposed Hereditary Properties of Diseases containing remarks on the unfounded terrors and ill-judged cautions consequent on such erroneous opinions* (London, 1814), p. 41.

4. '. . . to conceal the skeleton in the closet.' Ibid., preface.

5. '. . . a more accurate knowledge . . . suspense.' Ibid., p. 10.

6. '. . . unimportant to any family . . . free from them.' Ibid., p. 10.

7. '. . . so readily . . . not warranted.' Charles Davenport, 'Huntington's chorea in relation to heredity and eugenics', *American Journal of Insanity*, vol. 73 (1916), p. 217.

8. 'In marriage selection . . . two persons involved.' Ibid., p. 217.

9. 'We think only of personal liberty . . . liable to.' Ibid., p. 215.

10. 'Stupid.' Ibid., p. 215.

11. '. . . gave it me . . . decorously.' Sir Francis Galton, *Memories of My Life* (London, 1908), p. 290. Sir Francis tells this story against himself.

12. I apologise for any errors in this book.

13. '. . . mental deficiency.' *Hansard* (5th series, London, 1926), 29 November, column 967. Rhys Davies, MP.

14. Founded in 1907, the Eugenics Education Society changed its name in 1926, becoming the 'Eugenics Society'.

    The Society's present aims are: (a) to promote and support scientific research into the biological, genetic, social and cultural factors relating to human reproduction, development and health; (b) to evaluate and publish information on these matters for use by the medical and allied professions, for the enlightenment of decision-makers, and for the general education of the public; and (c) to foster a responsible attitude to parenthood, and to encourage circumstances in which the fullest achievement of individual human potential can be realised.

15. '. . . sterilisation of the "unfit" . . . spiritual wound.' J. A. Lindsay, *The Nineteenth Century and After*, vol. 72 (1912), p. 557.

16. 'Germany recently made . . . heritable conditions.' *Report of the 'Brock' Departmental Committee on Sterilisation* (London, 1934, Cmnd 4485), p. 34.

17. 'The right of society . . . State control.' Joseph Fletcher, *Morals and Medicine* (London, 1955), pp. 167–8.

18. 'I   of   give . . . family.' Adapted from the Edinburgh consent form which is set out in Alan E. H. Emery and D. L. Rimoin (eds), *Principles and Practice of Medical Genetics*, vol. 2, Edinburgh, (1983), p. 1,484.

19. The use of automatically processed information relating to individuals is governed by the Data Protection Act, 1984. Sections 21–25 of the Act give individuals a right to obtain details of personal data held about them and to ask the courts for compensation for damage suffered as a result of unauthorised destruction or disclosure, loss or inaccuracy.

    Section 29(1) of the Act empowers the Secretary of State to make an order which exempts information as to the physical or mental health of the data subject from the above access provisions; the Secretary of State is also empowered to modify the above provisions in relation to physical or mental health data.

    Note that the Act does not extend to a genetic register which is kept on cards in a shoe-box. For the purposes of the Act, 'data' means information recorded in a form in which it can be processed by equipment automatically in response to instructions given for that purpose. Some sections of the Act are not yet in force.

20. 'I felt I knew . . . pleased me.' *Personal communication.* In this case two doctors diagnosed very different conditions. The second opinion contradicted the first.

21. 'I tried to put him . . . 90 per cent okay.' A. Lippman-Hand and F. Clarke Fraser, 'Genetic counseling – the postcounseling period' in *American Journal of Medical Genetics,* vol. 4 (1979), p. 78.

22. 'I don't know . . . done?' Ibid., p. 61.

23. 'The doctor who delivered . . . again.' Ibid., p. 62.

24. 'The doctor told me not . . . might be either.' Anonymous 12-year-old girl cited by John Cohen, *Chance, Skill and Luck* (London, 1960), p. 36. The original quote has been added to and adapted for the purposes of this chapter.

25. 'The risk that any random . . . good risk.' J. A. Fraser Roberts, 'Genetics prognosis' in *British Medical Journal,* vol. 1, (1962), p. 589.

26. 'Many people . . . 1 in 200.' P. S. Harper, *Practical Genetic Counselling* (2nd edn, Bristol, 1984), p. 11.

27. 'Your chances of having a child . . . other couples . . . Before . . . 94 per cent.' Adapted from Thaddeus E. Kelly, *Clinical Genetics and Genetic Counseling* (Chicago and London, 1980), p. 293.

28. 'The chances of your potential grandchildren . . . their children.' *Personal communication.* Emphasis added.

29. See Abbey Lippman-Hand's and F. Clarke Fraser's 'Genetic Counseling: provision and reception of information' in *American Journal of Medical Genetics,* vol. 3 (1979). See also 'Postcounseling period' in *American Journal of Medical Genetics,* vol. 4 (1979), pp. 51–71 and pp. 73–87.

30. 'Your child would always need . . . please explain.' Adapted from J. R. Arnold and E. J. T. Winsor, 'The use of structured scenarios in genetic counselling' in *Clinical Genetics,* vol. 25 (1984), pp. 485–90.

31. 'A couple of years ago . . . needed his help.' Ruth Brandon, 'Explains her decision to abort a child' in *The Observer,* 12 June 1983.

32. 'If I hadn't already . . . different.' B. Shepperdson, 'Abortion and euthanasia of Down's syndrome children' in *Journal of Medical Ethics*, vol. 9 (1983), p. 153.

33. 'My husband and I . . . God sent.' Hilaire Gomer, 'On the dilemmas of delayed parenthood' in *The Times*, 1 Dec. 1983.

34. 'We decided not to go ahead . . . normal.' Ibid.

Glossary of Some Useful Terms

Abortion
Variously defined. Expulsion from the uterus of an embryo or fetus before the embryo or fetus is sufficiently mature to live outside the uterus.

Alleles
Alternative forms of a gene found at the same chromosomal locus.

Amino acid
Organic compound containing basic amino and acidic carboxyl groups.

Autosome
Chromosome that is not a sex chromosome.

Blastocyst
Cavity stage of early development, usually reached after the morula has entered the uterus.

Carrier
Variously defined. Person capable of transmitting a potentially harmful gene, even though he or she is unaffected. Sometimes used more widely.

Chromosome
Threadlike body in the nucleus that carries genetic information.

Cleavage
Splitting of the zygote by mitosis forms two daughter cells. Splitting of these cells forms four cells. And so, up to a point, mitotically on.

Conceptus
Embryo and its membranes.

Congenital
Present at birth (not necessarily genetic).

Consanguineous marriage
Marriage between persons who have at least one quite recent common ancestor. Often, in clinical practice, marriage between first cousins.

Crossing-over   Reciprocal exchange of genetic material between chromosomes.

Cytogenetics   Study of cellular features (principally chromosomes) in genetics.

Deletion   Chromosome aberration involving the loss of part of a chromosome.

Deoxyribonucleic acid (DNA)   Acid in chromosomes in which genetic information is coded. DNA is assisted by messenger ribonucleic acid (m-RNA) and transfer ribonucleic acid (t-RNA).

Dizygotic   Type of twins that arise when two separate eggs have been fertilised.

Dominant   Allele which determines the phenotype of persons who have an alternative form of the gene at the same locus.

Duplication   Chromosome aberration involving duplication of a chromosome.

Embryo   Very variously defined. The Warnock Committee of Inquiry into Human Fertilisation and Embryology regarded the six weeks immediately following fertilisation as the embryonic stage.

Eugenic   Of the production of fine offspring.

Fetus   The embryonic stage (see Embryo above) is followed by the fetal stage.

First-degree relatives   Parents, offspring and siblings.

Gamete   Reproductive cell, the ovum or sperm.

Gene   Variously defined. Unit of the material of inheritance, influencing a particular set of characters in a particular way. Part of the DNA molecule directing the synthesis of a specific polypeptide chain.

Genetic   Of or concerning genes. Of or concerning genetics.

Hemizygous male   Male expressing a trait determined by a gene on the X-chromosome.

Heterozygote	Person having two different alleles at one particular locus on a pair of homologous chromosomes.
Homologous chromosomes	Chromosomes containing identical sets of loci.
Homozygote	Person having two identical alleles at one particular locus on a pair of homologous chromosomes.
Hybrid	Cross-bred from two genetically different organisms.
Inversion	Chromosome aberration involving the turning upside down of a chromosome.
In vitro	In a glass.
Karotype	Number, size and shape of chromosomes in a somatic cell.
Linkage	Association of two or more genes caused by their location close to one another on the same chromosome.
Meiosis	Cell division resulting in gametes with half of the number of chromosomes found in the somatic cells.
Miscarriage	Variously defined. Interruption of pregnancy before the embryo or fetus is viable outside the uterus.
Mitosis	Cell division occurring in the somatic cells.
Monogenic	See Unifactorial below.
Monosomy	Loss of one of the members of a chromosome pair.
Monozygotic	Type of twins that arise when a single fertilised egg produces two viable individuals.
Morula	Mulberry-like sphere of embryonic cells prior to blastocyst formation.
Mosaicism (cellular)	Variations in the genetic composition of cells in an individual.
Multifactorial inheritance	Control of a character, not only by many genes, each of which has an additive effect, but also by the environment.

Mutation	A change in a gene or in the structure or number of chromosomes.
Neo-Darwinism	Modified theory of evolution, having no place for inheritance of acquired characteristics.
Nucleotides	Building blocks of nucleic acids.
Nucleus	Structure inside cells which contains chromosomes among other things.
Perinatal	Occurring round about birth.
Phenotype	All or any of the characteristics of a person.
Polypeptide	Organic compound comprising three or more amino acids.
Primordium	The earliest stage.
Pronucleus	Sperm nucleus after its entry into the ovum but before there has been fusion with the ovum nucleus. Ovum nucleus shortly before fusion with sperm nucleus.
Protein	Organic compound made of hundreds or thousands of amino acids.
Recessive	Allele which determines the phenotype of persons who are homozygous for the allele, but which does not determine the phenotypes of heterozygotes.
Ribonucleic acid (RNA)	See Deoxyribonucleic acid.
Risk	Exposure to mischance or peril. To endanger.
Segregation	Segregation of alleles during germ cell formation so that each sperm or egg contains only one member of each pair of alleles.
Sex chromosome	Chromosome responsible for the owner's sex.
Sibling	Sister or brother.
Somatic cells	Cells other than the germ cells.
Syndrome	Complex of symptoms and signs occurring together in particular disorders.
Termination	See Abortion.

Translocation   Transfer of genetic material, either to a non-homologous chromosome or to another part of the same chromosome.

Trimester   A period of three months. The nine calendar months of gestation are divided into three trimesters.

Trisomy   Gain of a chromosome.

Unifactorial   Inheritance controlled not by alleles at many different loci, but by a single gene pair.

X-linked   Traits determined by genes located on the X chromosome.

Zygote   Cell produced by the union of ovum and sperm and the fusion of their nuclei.

Bibliography

Abercrombie, M., *et al.* (eds), *The Penguin Dictionary of Biology* (London, 1980).

Adams, Joseph, *A Treatise on the Supposed Hereditary Properties of Diseases* (London, 1814).

Anon., 'Clinical genetics' (editorial) in *British Medical Journal*, vol. 287 (1983).

Anthony, E. J., and Benedek, T. (eds), *Parenthood: Its Psychology and Psychopathology* (Boston, 1970).

Antley, Mary Ann, *et al.*, 'Effects of genetic counselling on parental self-concepts' in *Journal of Psychology*, vol. 83 (1973).

Arbuthnot, John, 'An argument for divine providence' in *Philosophical Transactions of the Royal Society of London*, vol. 27 (1710).
Preface to *Of the Laws of Chance* (London, 1692).

Archbold, J. F., *Pleading, Evidence and Practice in Criminal Cases* (London, 1985).

Aristotle, *Generation of Animals*, trans. by A. L. Peck (London, 1943).
History of Animals, trans. by Richard Creswell (London, 1862).
The Politics, trans. by H. Rackham (London, 1932).
Aristotle's Compleat and Experienc'd Midwife, made in English by W– S– (London, 1700).

Arnold, J. R., and Winsor, E. J. T., 'The use of structured scenarios in genetic counselling' in *Clinical Genetics*, vol. 25 (1984).

Augustine, P. Charles, *A Commentary on the New Code of Canon Law*, vol. IV (London, 1920) and vol. VIII (London, 1922).

Austin, C. R., and Short, R. V. S. (eds), *Reproduction in Mammals: 1 Germ Cells and Fertilization* (Cambridge, 1982).

Reproduction in Mammals: 2 Embryonic and Fetal Development
   (Cambridge, 1982).
Ayer, A. J., *Probability and Evidence* (London, 1972).
Balinsky, B. I., and Fabian, B. C., *An Introduction to Embryology*
   (Philadelphia, 1981).
Baraitser, Michael, and Winter, Robin, *A Colour Atlas of Clinical Genetics*
   (London, 1983).
Barton, Mary, *et al.*, 'Artificial insemination' in *British Medical Journal*,
   vol. 1 (1945).
Bateson, Beatrice, *William Bateson F R S, Naturalist* (Cambridge, 1928).
Bateson, W., *The Methods and Scope of Genetics* (Cambridge, 1908).
Bible, Authorised King James Version (London, 1957).
Birch, Charles, and Abrecht, Paul (eds), *Genetics and the Quality of Life*
   (Potts Point, N S W and Oxford, 1975).
Blumberg, Bruce D., *et al.*, 'The psychological sequelae of abortion
   performed for a genetic indication' in *American Journal of Obstetrics
   and Gynecology*, vol. 122 (1975).
Boué Joëlle, *et al.*, 'Retrospective and prospective epidemiological studies
   of 1,500 karotyped human abortions' in *Teratology*, vol. 12 (1975).
Brambati, B., *et al.*, 'First trimester fetal diagnosis of genetic disorders:
   clinical evaluation of 250 cases' in *Journal of Medical Genetics*, vol. 22
   (1985).
Brandon, Ruth, 'Explains her decision to abort a child' in the *Observer*,
   12 June 1983.
British Agencies for Adoption and Fostering, *Adopting a Child: A Brief
   Guide for Prospective Adopters* (London, 1982).
British Medical Bulletin, 'Early prenatal diagnosis', vol. 39 (1983).
British Museum (Natural History), *Origin of Species* (Cambridge and
   London, 1981).
Brock (Chairman), *Report of the Departmental Committee on Sterilisation*
   (London, 1934, Cmnd 4485).
Brown, H. P., and Schanzer, S. N., *Female Sterilization* (Boston, 1982).
Callen, P. W., *Ultrasonography in Obstetrics and Gynecology* (Philadelphia,
   1983).
Calvin, Jean, *John Calvin: Documents of Modern History*, comp. by G. R.
   Potter and M. Greengrass (London, 1983).
Carlen, Claudia, *The Papal Encyclicals 1958–1981* (Washington D C,
   1981).
Carter, C. O., *Human Heredity* (London, 1962).
Carter, J., and Laurence, K. M., 'Genetic counselling during pregnancy

for couples at high risk for neural tube defect: is this the time to attend?' in *Biology and Society*, vol. 1 (1984).

Catholic Archbishops of Great Britain, *Abortion and the Right to Live* (Abbots Langley, 1980).

Chandrasekhar, S., *Abortion in a Crowded World* (London, 1974).

Chaucer, Geoffrey, *The Canterbury Tales*, ed. by A. C. Cawley (London, 1975).

Che'n, Jerome, *China and the West: Society and Culture, 1815–1937* (London, 1979).

Childs, B., 'Garrod, Galton and clinical medicine' in *Yale Journal of Biology and Medicine*, vol. 46 (1973).

Childs, B., *et al.*, 'Tay-Sachs' screening: motives for participating and knowledge of genetics and probability' and 'Tay-Sachs' screening: social and psychological impact' in *American Journal of Human Genetics*, vol. 28 (1976).

CIBA Foundation, *Human Genetics: Possibilities and Realities*, Symposium 66 (new ser.) (Amsterdam and Oxford, 1979).

Maternal Recognition of Pregnancy, Symposium 64 (new ser.) (Amsterdam and Oxford, 1979).

Clark, Ronald, *J. B. S. The Life and Work of J. B. S. Haldane* (London, 1968).

Clarke, Sir Cyril A., *Human Genetics and Medicine* (London, 1977).

Cohen, Jack, *Living Embryos* (Oxford, 1967).

Cohen, John, *Chance, Skill and Luck* (London, 1960).

Psychological Probability; or, The Art of Doubt (London, 1972).

Behaviour in Uncertainty and its Social Implications (London, 1964).

Coleman, William, *Biology in the Nineteenth Century: Problems of Form, Function and Transformation* (Cambridge, 1977).

Congenital Disabilities (Civil Liability) Act, 1976, *The Public and General Acts and General Synod Measures*, Part I (London, 1976).

Connor, J. M., *et al.*, 'Public awareness of genetic counselling services' in *Journal of Medical Genetics*, vol. 21 (1984).

Cooper, Thomas, *Certaine Sermons* (London, 1580).

Cooper, Wendy, and Smith, Tom, *Everything You Need to Know about the Pill* (London, 1984).

Côté, G. B., 'Odds in genetic counselling' in *Journal of Medical Genetics*, vol. 19 (1982).

Coxe, John Redman (trans.), *The Writings of Hippocrates and Galen. Epitomised from the Original Latin* (Philadelphia, 1846).

Dante Alighieri, *The Divine Comedy*, trans by H. W. Longfellow (London, 1890), Purgatory.

Darlington, C. D., *The Little Universe of Man* (London, 1978).

Darwin, Charles, *The Variation of Animals and Plants under Domestication* (London, 1875).

Darwin, Francis (ed.), *The Life and Letters of Charles Darwin Including an Autobiographical Chapter*, vol. 1 (London, 1887).

Data Protection Act, 1984 (London, 1984, C. 35).

Davenport, C. B., 'Reply to the criticism of recent American work by Dr Heron of the Galton Laboratory' in *Eugenics Record Office Bulletin no. 11* (Cold Spring Harbor, 1914).

 'Huntington's chorea in relation to heredity and eugenics, based on field notes made by Elizabeth Muncey, MD' in *American Journal of Insanity*, vol. 73 (1916).

Davern, Cedric I. (ed.), *Genetics* (*Readings from Scientific American*) (San Francisco, 1981).

Davis, Henry, *Artificial Human Fecundation* (London, 1951).

Davis, John A., 'Ethical issues in paediatric practice' in *Journal of the Royal Society of Medicine*, vol. 76 (1983).

Dawkins, Richard, *The Extended Phenotype* (Oxford and San Francisco, 1982).

Derham, W., *Physico-Theology: or, a Demonstration of the Being and Attributes of God, from his Works of Creation* (London, 1713).

Digby, Sir Kenelm, *Of Bodies, And of Mans Soul* (London, 1669).

Doi, A. Rahman I., *Introduction to the Hadith* (Sevenoaks, 1981).

Doncel, Joseph F., 'Immediate animation and delayed hominization' in *Theological Studies*, vol. 31 (1970).

Dunstan, G. R., 'The moral status of the human embryo: a tradition recalled' in *Journal of Medical Ethics*, vol. 1 (1984).

Eccles, Audrey, *Obstetrics and Gynaecology in Tudor and Stuart England* (London, 1982).

Edwards, A. W. F., *Likelihood* (London, 1976).

Edwards, R. G., and Purdy, J. M. (eds), *Human Conception* in vitro (London, 1982).

Edwards, Robert, and Steptoe, Patrick, *A Matter of Life* (London, 1980).

Eggleston, Richard, *Evidence, Proof and Probability* (London, 1978).

Emery, Alan E. H., *Elements of Medical Genetics* (Edinburgh, 1983).

 An Introduction to Recombinant DNA (Chichester, 1984).

 'The prevention of genetic disease in the population' in *International Journal of Environmental Studies*, vol. 3 (1972).

'A report on genetic registers' in *Journal of Medical Genetics*, vol. 15 (1978).

Emery, Alan E. H., and Rimoin, D. L. (eds), *Principles and Practice of Medical Genetics*, vols 1 and 2 (Edinburgh, 1983).

England, Marjorie A., *A Colour Atlas of Life Before Birth: Normal Foetal Development* (London, 1983).

Epstein, C. J., *et al.* (eds), *Risk, Communication and Decision-making in Genetic Counseling* (New York, 1979).

Erasmus, *Praise of Folly*, trans. by Betty Radice (London, 1971).

Eugenics Society, *Aims and Activities* (London, 1984).

Farley, John, *Gametes and Spores: Ideas About Sexual Reproduction 1750–1914* (London, 1982).

Feinberg, Joel, *The Problem of Abortion* (Belmont, 1973).

Feingold, M., and Pashayan, Hermione, *Genetics and Birth Defects in Clinical Practice* (Boston, 1983).

Festinger, Leon, *A Theory of Cognitive Dissonance* (London, 1959).

Feversham (Chairman), *Report on Human Artificial Insemination* (London, 1960, Cmnd 1105).

Fienus, Thomas, *De Formatrice Foetus* (Antwerp, 1620).

Fisher, R. A., 'The correlation between relatives on the supposition of Mendelian inheritance' in *Transactions of the Royal Society of Edinburgh*, vol. 52 (1918).

Fitzsimmons, J. S., 'The teaching of human genetics in schools' in *Journal of Medical Genetics*, vol. 20 (1983).

Fletcher, Joseph F., *Morals and Medicine* (London, 1955).
    The Ethics of Genetic Control, Ending Reproductive Roulette (New York, 1974)

Francke, Linda, *The Ambivalence of Abortion* (New York, 1978; London, 1979).

Freeden, Michael, 'Eugenics and progressive thought: a study in ideological affinity' in *The Historical Journal*, vol. 22 (1979).

Galton, Francis, *Hereditary Genius* (London, 1869).
    'The average contribution of each several ancestor to the total heritage of the offspring' in *Royal Society Proceedings* (*London*), vol. 61 (1897).
    Memories of My Life (London, 1908).

Garrod, A. E., 'The incidence of alkaptonuria' in *Lancet*, vol. II (1902).
    Inborn Errors of Metabolism (London, 1909).

George, Wilma, *Darwin* (London, 1982).

Gillies, D. A., *An Objective Theory of Probability* (London, 1973).

Gomer, Hilaire, 'On the dilemmas of delayed parenthood' in *The Times*, 1 December 1983.

Gorer, Geoffrey, *Exploring English Character* (London, 1955).

Graunt, John, *Natural and Political Observations mentioned in a following index, and made upon the Bills of Mortality* (London, 1662).

Gray, Muir, *Man Against Disease: Preventive Medicine* (Oxford, 1979).

Grisez, Germain, *Abortion: The Myths, the Realities and the Arguments* (New York, 1972).

Grobstein, Clifford, *From Chance to Purpose* (Reading, Mass., 1981).

Hacking, Ian, *The Emergence of Probability* (London, 1975).

Hall, A. R., *The Scientific Revolution 1500–1800* (London, 1954).

Hall, Robert E. (ed.), *Abortion in a Changing World*, vol. I (London, 1970).

Halsbury's Statutes of England, vol. 8 (London, 1969).

Handy, Joscelyn A., 'Psychological and social aspects of induced abortion' in *British Journal of Clinical Psychology*, vol. 21 (1982).

Hann, Judith, *The Perfect Baby?* (London, 1982).

Hansard, Parliamentary Debates: House of Commons (29 November 1926; 21 July 1931; and 7 March 1985).

Hanscombe, Gillian, 'The right to lesbian parenthood' in *Journal of Medical Ethics*, vol. 9 (1983).

Hardy, Alister, *The Biology of God* (London, 1975).

Harper, P. S., *Practical Genetic Counselling* (Bristol, 1981 and 1984).

Harrison, C. J., *et al.*, 'Investigation of human chromosome polymorphisms by scanning–electron microscopy' in *Journal of Medical Genetics*, vol. 22 (1985).

Hartsoeker, Nicolas, *Essay de Dioptrique* (Paris, 1694).

Hearnshaw, L. S., *Cyril Burt Psychologist* (London, 1979).

Heim, Alice, *Thicker than Water? Adoption: Its Loyalties, Pitfalls and Joys* (London, 1983).

Henslin, James M., 'Craps and magic' in *American Journal of Sociology*, vol. 73 (1967).

Heron, David, 'Mendelism and the problem of mental defect. I. A criticism of recent American work' in *Questions of the Day and the Fray* (no. VII, 1913).

Hilton, Bruce, and Callahan, Daniel (eds), *Ethical Issues in Human Genetics* (London, 1973).

Hodgson, Jane E., *Abortion and Sterilization: Medical and Social Aspects* (London and New York and San Francisco, 1981).

Horowitz, M. J., *Stress Response Syndromes* (London, 1976).

Hsia, Y. E., *et al.* (eds), *Counseling in Genetics* (New York, 1979).

Hsu, T. C., *Human and Mammalian Cytogenetics: An Historical Perspective* (New York, 1979).

Huff, Darrell, *How to Take a Chance* (London, 1965).

Human Concern, no. 18 (1985).

Huser, Roger J., *The Crime of Abortion in Canon Law* (Washington DC, 1942).

Iltis, Hugo, *Life of Mendel*, trans. by Eden and Cedar Paul (London, 1932).

Ison, David, *Artificial Insemination by Donor* (London, 1983).

Jacob, François, *The Possible and the Actual* (Seattle and London, 1982).

Janis, Irving L. (ed.), *Counseling on Personal Decisions: Theory and Research on Short-term Helping Relationships* (New Haven and London, 1982).

Janis, Irving L., and Mann, Leon, *Decision Making: A Psychological Analysis of Conflict, Choice and Commitment* (New York and London, 1977).

Jeffreys, Alec J., *et al.*, 'Positive identification of an immigration test-case using human DNA fingerprints' in *Nature*, vol. 317 (1985).

Jones, G. E., and Perry, C., 'Can claims for "wrongful life" be justified?' in *Journal of Medical Ethics*, vol. 9 (1983).

Keith, Sir Arthur, *Darwin Revalued* (London, 1955).

Kelly, Patricia T., *Dealing with Dilemma* (New York, 1977)

Kelly, Thaddeus E., *Clinical Genetics and Genetic Counseling* (Chicago and London, 1980).

Kessler, Seymour (ed.), *Genetic Counseling: Psychological Dimensions* (New York and London, 1979).

Kieffer, G. H., *Bioethics* (Reading, Mass., and London, 1979).

Kitzinger, Sheila, *Birth Over Thirty* (London, 1982).

Koran, trans. by N. J. Dawood (London, 1978)

Küng, Hans, *Infallible?* (London, 1971).

Lane (Chairman), *Report of the Committee on the Working of the Abortion Act* (London, 1974, Cmnd 5579).

Langley, D. V., *Making Decisions* (London, 1971).

Langley, Russell, *Practical Statistics: Simply Explained* (London, 1976).

Latt, S. A., and Darlington, G. J. (eds), 'Prenatal diagnosis: cell biological approaches' in *Methods in Cell Biology*, vol. 26 (1982).

Law Commission, *Family Law: Illegitimacy* (London, 1982; Law Com. no. 118).

Leff, Gordon, *Medieval Thought* (London, 1958).

Leonardo da Vinci, *The Notebooks of Leonardo da Vinci*, trans. and arranged by Edward MacCurdy (London, 1956).

Lewis, Abigail, *An Interesting Condition* (London, 1951).

Lindsay, James, 'The case for and against eugenics' in *The Nineteenth Century and After*, vol. 72 (1912).

Lippman-Hand, Abbey, and Fraser, F. Clarke, 'Genetic counseling: provision and reception of information' in *American Journal of Medical Genetics*, vol. 3 (1979).

'Genetic counseling – the postcounseling period' in *American Journal of Medical Genetics*, vol. 4 (1979).

Llewellyn-Jones, Derek, *Human Reproduction and Society* (London, 1974).

Locke, John, *An Essay Concerning Humane Understanding* (London, 1694).

Longford, Elizabeth, *Victoria R I* (London, 1964).

Lubs, H. A., and de la Cruz, F. (eds), *Genetic Counseling* (New York, 1977).

Lyons, Albert S., and Petrucelli, R. Joseph (eds), *Medicine: An Illustrated History* (New York, 1978).

McCormack, Michael K., *et al.*, 'Attitudes of persons at risk for Huntington's disease to reproduction by AID' in *American Journal of Medical Genetics*, vol. 14 (1983).

McCrae, W. M., *et al.*, 'Cystic fibrosis: parents' response to the genetic basis of the disease' in *Lancet*, vol. II (1973).

McKenzie, A. E., *The Major Achievements of Science*, vol. 1 (London, 1960).

McKusick, V. A., *Mendelian Inheritance in Man* (Baltimore and London, 1983).

Human Genetics (Englewood-Cliffs, NJ, 1969).

Mahoney, John, *Bioethics and Belief* (London, 1984).

Maistrov, L. E., *Probability Theory: A Historical Sketch*, trans. by Samuel Kotz (New York and London, 1974).

Markova, Ivana, *Paradigms, Thought and Language* (Chichester, 1982).

Mayr, Ernst, *The Growth of Biological Thought: Diversity, Evolution, and Inheritance* (Cambridge, Mass. and London, 1982).

Medawar, J. S., *The Life Science* (London, 1977).

Pluto's Republic (Oxford, 1982).

Mental Deficiency Bill, 1926 (no. 133) (HL) printed 15 July (London, 1926).

Mikkelsen, M., *et al.*, 'The impact of legal termination of pregnancy and prenatal diagnosis on the birth prevalence of Down's syndrome in Denmark' in *Annals of Human Genetics*, vol. 47 (1983).

Milunsky, Aubrey, *Know Your Genes* (London, 1980).

(ed.), *Genetic Disorders and the Fetus* (New York and London, 1979).

Milunsky, Aubrey, and Anas, George J. (eds), *Genetics and the Law* (2 vols, New York and London, 1976 and 1980).

Mohr, James C., *Abortion in America: The Origins and Evolution of National Policy, 1800–1900* (Oxford, 1978).

Moore, Keith L., *Before We Are Born: Basic Embryology and Birth Defects* (Philadelphia, 1983).

Moore, Peter G., *Reason By Numbers* (London, 1980).

Morris, Norman, and Arthure, Humphrey, *Sterilization as a Means of Birth Control in Men and Women* (London, 1976).

Mosteller, F., *et al.*, *Probability with Statistical Applications* (London, 1970).

Murphy, Edmund A., *Probability in Medicine* (London, 1979).

Nazer, Isam R. (ed.), *Induced Abortion: A Hazard to Public Health?* (Beirut, 1972).

Nicholas, Barry, *An Introduction to Roman Law* (Oxford, 1975).

Nicholson, Susan Teft, *Abortion and the Roman Catholic Church* (Knoxville, Tenn., 1978).

Nolte, Ernst, *Three Faces of Fascism* (London, 1965).

Noonan, John T. Jr, *Contraception. A History of Its Treatment by the Catholic Theologians and Canonists* (Cambridge, Mass., 1966).

   (ed.), *The Morality of Abortion: Legal and Historical Perspectives* (Cambridge, Mass., 1970).

Orel, V., *Mendel* (Oxford, 1984).

Paré, Ambroise, *The Works of Ambrose Parey. Translated out of Latine and compared with the French* by Thomas Johnson (London, 1634).

Payne, E. C., *et al.*, 'Outcome following therapeutic abortion' in *Archives of General Psychiatry*, vol. 33 (1976).

Pearce, Fred, 'United Nations Conference Rewards Contraceptive Technology' in *New Scientist*, 16 Aug. 1984.

Pearn, J. H., 'Patients' subjective interpretation of risks offered in genetic counselling' in *Journal of Medical Genetics*, vol. 10 (1973).

'Peel Report on Human Artificial Insemination' in *British Medical Journal*, vol. II (1973).

Persaud, T. V. N. (ed.), *Genetic Disorders, Syndromology and Prenatal Diagnosis* (Lancaster, 1982).

Perutz, Max, 'Cancer: causes, cures and the molecular biologist' in *The Times Higher Education Supplement*, 4 March 1983.

Pfeffer, Naomi, and Woollett, Anne, *The Experience of Infertility* (London, 1983).

Plato, *The Republic* trans. by A. D. Lindsay (London, 1923).

Popper, Karl R., *Realism and the Aim of Science* from the *Postscript* to *The Logic of Scientific Discovery*, ed. by W. W. Bartley III (London, 1983).

Port Royal 'Art of thinking' in *Course of Education Pursued at the Universities of Cambridge and Oxford*, vol. 3 (London, 1818).

Porter, Ian H., and Hook, Ernest B. (eds), *Service and Education in Medical Genetics* (New York and London, 1979).

Potts, Malcolm, *et al.*, *Abortion* (Cambridge, 1977).

Ramaswamy, Saroja, and Smith, Tony, *Practical Contraception* (Tunbridge Wells, 1976).

Reading, A. E., *et al.*, 'Health beliefs and health care behaviour in pregnancy' in *Psychological Medicine*, vol. 12 (1982).

Roberts, J. A. Fraser, 'Genetic prognosis' in *British Medical Journal*, vol. I (1962).

Roberts, J. A. Fraser, and Pembrey, Marcus E., *An Introduction to Medical Genetics* (Oxford, 1978).

Rodeck, C. H., and Nicolaides, K. H. (eds), *Prenatal Diagnosis – Proceedings of the Eleventh Study Group of the RCOG* (London, 1984).

Rogers, W. V. H., *Winfield and Jolowicz on Tort* (London, 1984).

Romer, Alfred Sherwood, and Parsons, Thomas S., *The Vertebrate Body* (London, 1977).

Rothwell, Norman V., *Understanding Genetics* (New York, 1983).

Rowland, Beryl, *Medieval Woman's Guide to Health: The First English Gynecological Handbook* (London, 1981).

Royal College of Obstetricians and Gynaecologists Party, 'Report on ultrasound examination in pregnancy' (London, 1984).

Royal Society Proceedings, 'The assessment and perception of risk', ser. A, vol. 376, no. 1764, 30 April (London, 1981).

Ruestow, Edward G., 'Images and ideas: Leeuwenhoek's perception of the spermatozoa' in *Journal of the History of Biology*, vol. 16 (1983).

Russell, Colin, *Science and Social Change (1700–1900)* (London, 1983).

Sandler, Iris, 'Pierre Louis Moreau de Maupertius – a precursor of Mendel?' in *Journal of the History of Biology*, vol. 16 (1983).

Sansom, G. B., *Japan. A Short Cultural History* (London, 1976).

Searle, G. R., *Science in History Number 3: Eugenics and Politics in Britain, 1900–1914* (Leyden, 1976).

Session Cases (Edinburgh, 1958).

Shepperdson, Billie, 'Abortion and euthanasia of Down's syndrome children – the parents' view' in *Journal of Medical Ethics*, vol. 9 (1983).

Shussman, Leon N., and Schatkin, Sidney B., 'Blood-grouping tests in undisputed paternity proceedings' in *American Medical Association*, vol. 164 (1957).

Siegler, Mark, 'Pascal's Wager and the Hanging of Crepe' in *New England Journal of Medicine*, vol. 293 (1975).

Siggers, D. C., *Prenatal Diagnosis of Genetic Disease* (Oxford, 1978).

Singer, Charles, *A History of Biology* (London, 1959).

Singer, Peter, *Practical Ethics* (Cambridge, 1979).

Smith, Anthony, *The Human Pedigree* (London, 1975).

Smith, C. P. Wendall, *et al.*, *Basic Human Embryology* (London, 1984).

Smith, J. C., and Hogan, Brian, *Criminal Law* (London, 1978 and 1983).

Snowden, R., and Mitchell, G. D., *The Artificial Family* (London, 1981).

Snowden, R., *et al.*, *Artificial Reproduction* (London, 1983).

Soranus, *Gynecology*, trans. by Owsei Temkin (Baltimore, 1956).

Sterne, Laurence, *Tristram Shandy*, vol. 1 (London, 1760).

Stevenson, Alan Carruth, and Davison, B. C. Clare, *Genetic Counselling* (London, 1976).

Surrogacy Arrangements Act, 1985 (London, 1985, C. 49).

Sutton, H. Eldon, *An Introduction to Human Genetics* (Philadelphia, 1980).

Swallerstein, J., *et al.*, 'Psychological sequelae of therapeutic abortion in young unmarried women' in *Archives of General Psychiatry*, vol. 27 (1972).

Swanson, Carl P., *et al.*, *Cytogenetics: The Chromosome in Division, Inheritance and Evolution* (Englewood-Cliffs, NJ, 1981).

Talmud (Babylonian), trans. under editorship of D. R. I. Epstein, Yebamoth I (London, 1936).

Targum, Steven D., 'Psychotherapeutic considerations in genetic counseling' in *American Journal of Medical Genetics*, vol. 8 (1981).

Taussig, Frederick J., *Abortion Spontaneous and Induced: Medical and Social Aspects* (London, 1936).

Taylor, Gordon Rattray, *The Science of Life* (London, 1963).

Thomas, Keith, *Man and the Natural World* (London, 1983).

   Religion and the Decline of Magic (London, 1978).

Thompson, James S., and Thompson, Margaret W., *Genetics in Medicine* (Philadelphia, 1980).

Thomson, W. A. R., *A Dictionary of Medical Ethics and Practice* (Bristol, 1977).

Tietze, Christopher, and Lewit, Sarah, 'Abortion' in *Scientific American* (new ser.), vol. 220 (1969).

Titus Lucretius Carus, *Titus Lucretius Carus on the Nature of Things*, trans. by Thomas Jackson (Oxford, 1929).

Treasure, G. R. R., *Seventeenth-century France* (London, 1966).

Uebele-Kallhardt, B. M., *Human Oocytes and their Chromosomes: An Atlas* (Berlin, 1978).

United States Reports (Washington, DC, 1973).

Varley, H. Paul, *Japanese Culture* (London, 1973).

Vogel, F., and Motulsky, A. G., *Human Genetics: Problems and Approaches* (Berlin and Heidelberg and New York, 1979).

Walters, William, and Singer, Peter, *Test-Tube Babies: A Guide to Moral Questions, Present Techniques and Future Possibilities* (Oxford, 1982).

Warnock (Chairman), *Report on Human Fertilisation and Embryology* (London, 1984, Cmnd 9314).

Watson, James D., *The Double Helix* (London, 1968).

Weatherall, D. J., *The New Genetics and Clinical Practice* (London, 1982).

Wellings, Kaye, and Mills, Angela, 'Contraceptive trends' in *British Medical Journal*, vol. 289 (1984).

Wertz, Dorothy C., *et al.*, 'Genetic counseling and reproductive uncertainty' in *American Journal of Medical Genetics*, vol. 18 (1984).

Williams, Glanville, 'The law of abortion' in *Current Legal Problems*, vol. v (1952).

  The Sanctity of Life and the Criminal Law (London, 1958).

Winchester, A. M., *Human Genetics* (Columbus and London, 1979).

Zuk, G. H., 'The religious factors and the role of guilt in parental acceptance of the retarded child' in *American Journal of Mental Deficiency*, vol. 64 (1959).

Index

MORE ABOUT PENGUINS, PELICANS, PEREGRINES AND PUFFINS

For further information about books available from Penguins please write to Dept EP, Penguin Books Ltd, Harmondsworth, Middlesex UB7 0DA.

In the U.S.A.: For a complete list of books available from Penguins in the United States write to Dept DG, Penguin Books, 299 Murray Hill Parkway, East Rutherford, New Jersey 07073.

In Canada: For a complete list of books available from Penguins in Canada write to Penguin Books Canada Limited, 2801 John Street, Markham, Ontario L3R 1B4.

In Australia: For a complete list of books available from Penguins in Australia write to the Marketing Department, Penguin Books Australia Ltd, P.O. Box 257, Ringwood, Victoria 3134.

In New Zealand: For a complete list of books available from Penguins in New Zealand write to the Marketing Department, Penguin Books (N.Z.) Ltd, Private Bag, Takapuna, Auckland 9.

In India: For a complete list of books available from Penguins in India write to Penguin Overseas Ltd, 706 Eros Apartments, 56 Nehru Place, New Delhi 110019.

CASTAWAY   *Lucy Irvine*

'Writer seeks "wife" for a year on a tropical island.' This is the extraordinary, candid, sometimes shocking account of what happened when Lucy Irvine answered an advertisement, and found herself embroiled in what was not exactly a desert island dream. 'Fascinating' – *Daily Mail*

OUT OF AFRICA   *Karen Blixen (Isak Dinesen)*

After the failure of her coffee-farm in Kenya, where she lived from 1913 to 1931, Karen Blixen went home to Denmark and wrote this unforgettable account of her experiences. 'No reader can put the book down without some share in the author's poignant farewell to her farm' – *Observer*

THE LISLE LETTERS
Edited by Muriel St Clare Byrne

An intimate, immediate and wholly fascinating picture of a family in the reign of Henry VIII. 'Remarkable ... we can really hear the people of early Tudor England talking' – Keith Thomas in the *Sunday Times*. 'One of the most extraordinary works to be published this century' – J. H. Plumb

IN MY WILDEST DREAMS   *Leslie Thomas*

The autobiography of Leslie Thomas, author of *The Magic Army* and *The Dearest and the Best:* From Barnardo boy to original virgin soldier, from apprentice journalist to famous novelist, it is an amazing story. 'Hugely enjoyable' – *Daily Express*

INDIA: THE SIEGE WITHIN   *M. J. Akbar*

'A thoughtful and well-researched history of the conflict, 2,500 years old, between centralizing and separatist forces in the sub-continent. And remarkably, for a work of this kind, it's concise, elegantly written and entertaining' – Zareer Masani in the *New Statesman*

THE WINNING STREAK
Walter Goldsmith and David Clutterbuck

Marks and Spencer, Saatchi and Saatchi, United Biscuits, G.E.C. ... The U.K's top companies reveal their formulas for success, in an important and stimulating book that no British manager can afford to ignore.

ADIEUX: A FAREWELL TO SARTRE
Simone de Beauvoir

A devastatingly frank account of the last years of Sartre's life, and his death, by the woman who for more than half a century shared that life. 'A true labour of love, there is about it a touching sadness, a mingling of the personal with the impersonal and timeless which Sartre himself would surely have liked and understood' – *Listener*

BUSINESS WARGAMES *James Barrie*

How did B M W overtake Mercedes? Why did Laker crash? How did McDonalds grab the hamburger market? Drawing on the tragic mistakes and brilliant victories of military history, this remarkable book draws countless fascinating parallels with case histories from industry worldwide.

METAMAGICAL THEMAS *Douglas R. Hofstadter*

This astonishing sequel to the best-selling, Pulitzer Prize-winning *Godel, Escher, Bach* swarms with 'extraordinary ideas, brilliant fables, deep philosophical questions and Carrollian word play' – Martin Gardner

INTO THE HEART OF BORNEO *Redmond O'Hanlon*

'Perceptive, hilarious and at the same time a serious natural-history journey into one of the last remaining unspoilt paradises' – *New Statesman*. 'Consistently exciting, often funny and erudite without ever being overwhelming' – *Punch*

A BETTER CLASS OF PERSON *John Osborne*

The playwright's autobiography, 1929–56. 'Splendidly enjoyable' – John Mortimer. 'One of the best, richest and most bitterly truthful autobiographies that I have ever read' – Melvyn Bragg

THE SECRETS OF A WOMAN'S HEART
Hilary Spurling

The later life of Ivy Compton-Burnett, 1920–69. 'A biographical triumph . . . elegant, stylish, witty, tender, immensely acute – dazzles and exhilarates . . . a great achievement' – Kay Dick in the *Literary Review*. 'One of the most important literary biographies of the century' – *New Statesman*

THE APARTHEID HANDBOOK   Roger Omond

A guide to South Africa's everyday racial policies. Through direct questions and answers, the *Handbook* fills the gaps left by much conventional reporting and gives us the hard facts – both the theory behind apartheid and the practice.

JEAN RHYS: LETTERS 1931–66
Edited by Francis Wyndham and Diana Melly

'Eloquent and invaluable . . . her life emerges, and with it a portrait of an unexpectedly indomitable figure' – Marina Warner in the *Sunday Times*

AMONG THE RUSSIANS   Colin Thubron

One man's solitary journey by car across Russia provides an enthralling and revealing account of the habits and idiosyncrasies of a fascinating people. 'He sees things with the freshness of an innocent and the erudition of a scholar' – *Daily Telegraph*

THE AMATEUR NATURALIST
Gerald Durrell with Lee Durrell

'Delight . . . on every page . . . packed with authoritative writing, learning without pomposity . . . it represents a real bargain' – *The Times Educational Supplement*. 'What treats are in store for the average British household' – *Books and Bookmen*

MRS THATCHER'S ECONOMIC EXPERIMENT
William Keegan

Harsh realities, forthright historical analysis and a wealth of anecdotes combine to make this a highly stimulating account of an experiment that has dramatically failed. 'A short but powerful and devastating book' – *Listener*

THEY WENT TO PORTUGAL   Rose Macaulay

An exotic and entertaining account of travellers to Portugal from the pirate-crusaders, through poets, aesthetes and ambassadors, to the new wave of romantic travellers. A wonderful mixture of literature, history and adventure, by one of our most stylish and seductive writers.

THE BOOK QUIZ BOOK *Joseph Connolly*

Who was literature's performing flea . . .? Who wrote 'Live Now, Pay Later . . .'? Keats and Cartland, Balzac and Braine, Coleridge conundrums, Eliot enigmas, Tolstoy teasers . . . all in this brilliant quiz book. You will be on the shelf without it . . .

VOYAGE THROUGH THE ANTARCTIC
Richard Adams and Ronald Lockley

Here is the true, authentic Antarctic of today, brought vividly to life by Richard Adams, author of *Watership Down*, and Ronald Lockley, the world-famous naturalist. 'A good adventure story, with a lot of information and a deal of enthusiasm for Antarctica and its animals' – *Nature*

GETTING TO KNOW THE GENERAL *Graham Greene*

'In August 1981 my bag was packed for my fifth visit to Panama when the news came to me over the telephone of the death of General Omar Torrijos Herrera, my friend and host . . .' 'Vigorous, deeply felt, at times funny, and for Greene surprisingly frank' – *Sunday Times*

TELEVISION TODAY AND TOMORROW: WALL TO WALL DALLAS?
Christopher Dunkley

Virtually every British home has a television, nearly half now have two sets or more, and we are promised that before the end of the century there will be a vast expansion of television delivered via cable and satellite. How did television come to be so central to our lives? Is British television really the best in the world, as politicians like to assert?

ARABIAN SANDS *Wilfred Thesiger*

'In the tradition of Burton, Doughty, Lawrence, Philby and Thomas, it is, very likely, the book about Arabia to end all books about Arabia' – *Daily Telegraph*

WHEN THE WIND BLOWS *Raymond Briggs*

'A visual parable against nuclear war: all the more chilling for being in the form of a strip cartoon' – *Sunday Times*. 'The most eloquent anti-Bomb statement you are likely to read' – *Daily Mail*

A FORTUNATE GRANDCHILD *'Miss Read'*

Grandma Read in Lewisham and Grandma Shafe in Walton-on-the-Naze were totally different in appearance and outlook, but united in their affection for their grand-daughter – who grew up to become the much-loved and popular novelist.

THE ULTIMATE TRIVIA QUIZ GAME BOOK
Maureen and Alan Hiron

If you are immersed in trivia, addicted to quiz games, endlessly nosey, then this is the book for you: over 10,000 pieces of utterly dispensable information!

THE DIARY OF VIRGINIA WOOLF
Five volumes, edited by Quentin Bell and Anne Olivier Bell

'As an account of the intellectual and cultural life of our century, Virginia Woolf's diaries are invaluable; as the record of one bruised and unquiet mind, they are unique' – Peter Ackroyd in the *Sunday Times*

VOICES OF THE OLD SEA *Norman Lewis*

'I will wager that *Voices of the Old Sea* will be a classic in the literature about Spain' – *Mail on Sunday*. 'Limpidly and lovingly Norman Lewis has caught the helpless, unwitting, often foolish, but always hopeful village in its dying summers, and saved the tragedy with sublime comedy' – *Observer*

THE FIRST WORLD WAR *A. J. P. Taylor*

In this superb illustrated history, A. J. P. Taylor 'manages to say almost everything that is important for an understanding and, indeed, intellectual digestion of that vast event ... A special text ... a remarkable collection of photographs' – *Observer*

NINETY-TWO DAYS *Evelyn Waugh*

With characteristic honesty, Evelyn Waugh here debunks the romantic notions attached to rough travelling: his journey in Guiana and Brazil is difficult, dangerous and extremely uncomfortable, and his account of it is witty and unquestionably compelling.

A choice of Penguins and Pelicans

THE BIG RED TRAIN RIDE *Eric Newby*

From Moscow to the Pacific on the Trans-Siberian Railway is an eight-day journey of nearly six thousand miles through seven time zones. In 1977 Eric Newby set out with his wife, an official guide and a photographer on this journey. 'The best kind of travel book' – Paul Theroux

STAR WARS *Edited by E. P. Thompson*

With contributions from Rip Bulkeley, John Pike, Ben Thompson and E. P. Thompson, and with a Foreward by Dorothy Hodgkin, OM, this is a major book which assesses all the arguments for Star Wars and proceeds to make a powerful – indeed unanswerable – case against it.

SELECTED LETTERS OF MALCOLM LOWRY
Edited by Harvey Breit and Margerie Bonner Lowry

Lowry emerges from these letters not only as an extremely interesting man, but also a lovable one' – Philip Toynbee

PENGUIN CLASSICS OF WORLD ART

Each volume presents the complete paintings of the artist and includes: an introduction by a distinguished art historian, critical comments on the painter from his own time to the present day, 64 pages of full-colour plates, a chronological survey of his life and work, a basic bibliography, a fully illustrated and annotated *catalogue raisonne*.

TITLES ALREADY PUBLISHED OR IN PREPARATION

Botticelli, Bruegel, Canaletto, Caravaggio, Cezanne, Durer, Giorgione, Giotto, Leonardo da Vinci, Manet, Mantegna, Michelangelo, Picasso, Piero della Francesca, Raphael, Rembrandt, Toulouse-Lautrec, van Eyck, Vermeer, Watteau

A choice of Penguin and Pelicans

ASIMOV'S NEW GUIDE TO SCIENCE
Isaac Asimov

A fully updated edition of a classic work – far and away the best one-volume survey of all the physical and biological sciences.

CHILDREN AND COMPUTERS
Martin Hughes and Hamish MacLeod

A major study of a crucial issue: how computers will affect the emotional and mental development of all our children.

THE DOUBLE HELIX *James D. Watson*

Watson's vivid and outspoken account of how he and Crick discovered the structure of DNA (and won themselves a Nobel Prize) – one of the greatest scientific achievements of the century.

EVER SINCE DARWIN *Stephen Jay Gould*

'Stephen Gould's writing is elegant, erudite, witty, coherent and forceful' – Richard Dawkins, *Nature*

MATHEMATICAL MAGIC SHOW *Martin Gardner*

A further mind-bending collection of puzzles, games and diversions by the undisputed master of recreational mathematics.

SILENT SPRING *Rachel Carson*

The brilliant book which provided the impetus for the ecological movement – and has retained its supreme power to this day.

A choice of Penguins and Pelicans

THE SOUL OF A NEW MACHINE *Tracy Kidder*

The gripping story of a team of young computer wizards and their superhuman efforts to achieve a technological breakthrough, this enthralling book won the Pulitzer Prize for 1982.

BRIGHTER THAN A THOUSAND SUNS
Robert Jungk

'By far the most interesting historical work on the atomic bomb I know of' – C. P. Snow

TURING'S MAN *J. David Bolter*

We live today in a computer age, which has meant some startling changes in the ways we understand freedom, creativity and language. This major book looks at the implications.

EINSTEIN'S UNIVERSE *Nigel Calder*

'A valuable contribution to the de-mystification of relativity' – *Nature*

THE CREATIVE COMPUTER
Donald R. Michie and Rory Johnston

Computers *can* create the new knowledge we need to solve some of our most pressing human problems; this path-breaking book shows how.

ONLY ONE EARTH *Barbara Ward and Rene Dubos*

An extraordinary document which explains with eloquence and passion how we should go about 'the care and maintenance of a small planet'.